T0187143

# MANY

# THINGS

# UNDER

# A ROCK

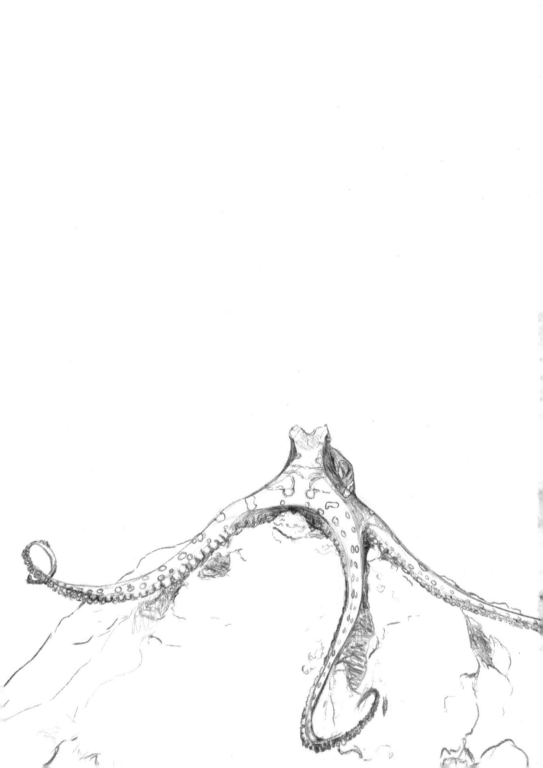

# MANY THINGS UNDER A ROCK

## The Mysteries of Octopuses

## DAVID SCHEEL

Illustrations by Laurel "Yoyo" Scheel

**HODDER &
STOUGHTON**

First publishing in the US in 2023 by W. W. Norton & Company
First published in Great Britain in 2023 by Hodder & Stoughton
An Hachette UK company

2

Copyright © David Scheel 2023

The right of David Scheel to be identified as the Author of the Work has been asserted
by him in accordance with the Copyright, Designs and Patents Act 1988.

Illustrations copyright © Laurel Scheel

All rights reserved. No part of this publication may be reproduced, stored in a retrieval system,
or transmitted, in any form or by any means without the prior written permission of the
publisher, nor be otherwise circulated in any form of binding or cover other than that in which
it is published and without a similar condition being imposed on the subsequent purchaser.

A CIP catalogue record for this title is available from the British Library

Hardback ISBN 978 1 529 39260 9
Trade Paperback ISBN 978 1 529 39261 6
eBook ISBN 978 1 529 39262 3

Printed and bound in Great Britain by Clays Ltd, Elcograf S.p.A.

Hodder & Stoughton policy is to use papers that are natural, renewable and recyclable
products and made from wood grown in sustainable forests. The logging and manufacturing
processes are expected to conform to the environmental regulations of the country of origin.

Hodder & Stoughton Ltd
Carmelite House
50 Victoria Embankment
London EC4Y 0DZ

www.hodder.co.uk

*For Peyton and Robyn, who are always with me.*

# CONTENTS

## 3: REACH

### Sensation and the Grasp of Octopuses

### Octopus Cognition

## 4: REVELATION

### Solitary Octopuses

### Society Octopuses

# MANY

# THINGS

# UNDER

# A ROCK

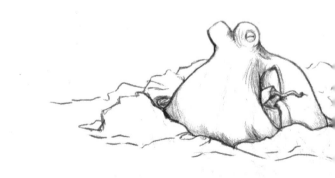

# INTRODUCTION

## The Inner Lives of Octopuses

### Underwater in Prince William Sound, Southcentral Alaska

I followed the octopus into the emerald void between the surface and the depths. Above us, in sunlight, the water sparkled with tiny, glassy plankton of some sort; below us the light vanished in darkening green over the flats just out of sight.

I swam parallel with the octopus, finning hard to match his pace. The octopus was a large juvenile male of perhaps twenty pounds, and missing his second right arm. He planed backward through the water—mantle first, then eyes and head, arms trailing. His web was flattened into a hydrofoil, web membrane extended, stump and three arms above, four arms below. His mantle expanded and then contracted powerfully, expelling the water from his siphon as he jetted over the distant, silted bottom.

The octopus pitched his body downward, and we descended. He neared the bottom at a depth of about forty feet, and abruptly blanched white. He spread his arms and web wide, then grabbed hold of some kelp. White phantasms raced out of his blanched skin, chased by incoming waves of burnt and red ochre. In a second, he matched the surrounding Laminariales kelp. He pulled a broad kelp blade over his head.

"I have disappeared," he seemed to say.

The kelp blade covered his eyes, head, and much of his central body. Yet he was still quite apparent—his arms and the back of his mantle remained outside the umbra of the kelp. I lowered myself as close to the bottom as I could in my bulky dive gear, pressed my cheek down, and peeked under his chosen shroud. A black horizontal pupil gazed back, lid slightly lowered, papillae raised in horns over each eye. On noting me, one arm curled across the eye as though he didn't want me to see him, hiding there, but not very well.

At that time, I had barely begun training as a scientific diver, and I was new to octopuses. I left him then, as I headed up to the surface, to air, light, and warmth above. I was curious about everything I had seen, from the backward and flattened swimming posture, to the wash of colors over the skin, the missing limb, and perhaps most of all, that hidden eye, peering from under the kelp, peering out from behind his own suckers.

### Underwater off Cannikin and Long Shot Nuclear Test Sites, Amchitka Island, Aleutians, Alaska

GIANT PREDATORY BEASTS live in the cold, remote depths of the Bering Sea. One of these was watching me.

Even in July, the waters around Amchitka were only 41 degrees Fahrenheit—I was glad of my dry suit keeping me warm. My view was dominated by canyon walls plastered up to ten inches thick with hard, calcareous algae in shades of amaranth and baby pink. Green sea urchins dotted the narrow seafloor between, and kept it clear of any soft algae vulnerable to their appetites. Sponges jutted out from the concreted walls and canyon floor in stalks and balls and towers, translucent or colored red or pale ivory. Clusters of horse mussels, each longer than my hand, grew in cracks in the pavement.

Stands of tightly bunched *Alaria* kelp rose here and there, small blades near the bottom sweeping around their holdfasts—possibly to discourage grazing urchins—and long stipes leading up to the floats and larger fronds at the surface. Everything that wasn't rigidly attached swayed with the surge.

Thirty-five feet above me, an Achilles rigid inflatable skiff tossed on the waves beneath mist-shrouded pinnacles and cliffs of the Aleutian landscape. I had fallen out of that skiff, a thousand miles from my home in Anchorage, two miles from our mothership (the research vessel *Norseman*), into the Seussian-canyoned landscape submerged below.

I was in unfamiliar territory.

It was my second day in the Aleutians, fifteen years after my earliest diving in Prince William Sound. My scuba buddy had just completed scientific diver training, and we were still learning to dive together. We were both beginners here in different ways, working with a team of elite scientific divers from Alaska's premier university, and in one of the state's most remote and challenging environments. We were here to address a simple question—was the local marine environment radioactive?

The United States had detonated three nuclear bombs at Amchitka Island in the 1960s and early 1970s during underground testing. The final five megaton explosion, buried over a mile below the surface, had caused a seismic shock of nearly 7.0 magnitude, about the same magnitude that cracked and tilted my garage floor, and staggered me in the kitchen of my Anchorage home during a December 2018 earthquake. Was any contamination leaking through cracks in the bedrock out of those deep test holes from forty years ago, to bioaccumulate in the kelp, urchins, and predators of the Aleutian marine ecosystem?

I looked up overhead as my buddy approached. We had earlier moored a bag above me on the anchor line to bring to the surface, stuffed to the size of Santa's sack with kelp collected during the first part of the dive. Tugged by the bag and line, the anchor banged and scraped its hold on the hard bottom as the boat tossed in the chop above.

We were nearing the end of the dive. I checked my air supply—500 psi—yes, time to surface.

Then, my dive buddy pointed over my shoulder, forefinger stitching into the water for emphasis. Her eyes were wide, bubbles clattering in alarm from her scuba regulator. I turned to look.

The head of a large octopus was rising slowly up over a nearby ridge of rock and pink coralline algae. The thick first left arm slithered over the ridgeline in a broad curl, descending down toward us and the anchor. Behind the advancing arm, the head of the octopus rose ever higher. I wondered momentarily if it was stalking us.

How big was this animal, anyway—she looked huge and appeared to be growing larger as each additional sucker was revealed. "Holy s—!" my dive buddy later yelled to the dive tenders, spitting out her regulator when we did reach the surface. It was her first octopus encounter as a diver. "This octopus is huge!"

The anchor clattered on the rocks some more. We were ringing a gong, and an ocean giant had answered.

"We don't find octopuses very often," I had earlier been told about diving in the Aleutians. "But when we do, they're big! The octopuses come to investigate the sound of the anchor clanking on the hard bottom." This story was a bit surprising—at the time, the state of science about octopus hearing was that they were "probably not deaf." What they could hear, how (what organ functioned as ears?), and why, were questions that had not, by then, yet received definitive study. We now have a better understanding that octopuses, and their relatives, the cuttlefish and squid, do respond to some ocean sounds, and that these are detected by organs that, like the vestibular system in our inner ears, also help to regulate balance and to sense changes in movement.

WHEN I BEGAN MY CAREER, octopuses already were recognized as solitary animals, indifferent to each other except as threat or prey—cannibals, with limited means to recognize their fellow beings.

My curiosity was as strong as ever. Had that first octopus, years ago, been embarrassed to be found? Octopuses are supposed to be good at hiding. Could an octopus feel embarrassed?

What interest could this second octopus, out in the remote Aleutians, have in the clangor of metal on rock? What source of food or mates made a noise like that? I could think of none. What was really going on, when a giant octopus answered the call of a clanging anchor to stalk the divers working there?

THESE TWO ENCOUNTERS, and the many questions that occurred to me in relation to them, exemplified my challenges over twenty-five years studying octopuses. I was curious to know how octopuses live and experience the world. They cannot tell us how they feel, however, and they are worlds different from humans and other mammals. To some, these were reasons enough to feel that it was not possible to learn about the inner lives of octopuses. What can we know, even in principle, about what it is like to be an octopus?

That picture is changing. Over my professional lifetime, people have built, discovered, and learned new scientific ways to understand the experiences of animals. These have led us to discover previously unknown aspects of octopus physiology, and previously unknown species, habitats, and behaviors. As these new ways of study develop, so also does our interest in what once might have been merely rhetorical questions.

One route to knowing others is also the most obvious. Behavior, in particular, reveals values—what is good and what is bad to an octopus—and hints at an animal's experiences and intentions.

In *Many Things Under a Rock*, we will try to find some answers. Or, at least, try to see more clearly the shape that these answers must have so that we can better understand the inner lives of octopuses. This is a story of what we have learned and what we are still learning about the natural history and lives of octopuses.

# 1

## Where Are They?

# *Missing Octopuses*

# 1

## Starting Out in Alaska

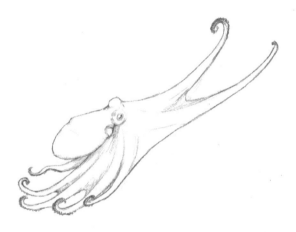

I met my first octopus in Cordova, in southeastern Alaska, in the new lobby aquarium in the US Forest Service building. She had arrived there in a bucket, brought by the fisherman who pulled her, entangled with a kelp holdfast, from the muddy Copper River Delta to the bow-picker deck.

In the lobby aquarium tank, she was a glistening Christmas tree ornament of curling arms lined with bone china suckers, tea-brown skin flecked with clotted cream, and black pupils whose narrow views admitted her world. She bobbed her head up and down at my peering round face. Without any change of her attention, two of her rear arms sought the back of the aquarium and found a gap under a rock. These two were followed by another arm, and soon poured the rest of her into the same gap, where her eyes still looked out. In a moment, the eyes also disappeared into shadow. The visible space of the tank was empty.

I came back on several other days to visit and look into the tank. She was given the name Ophelia—a phonetic pun. I never saw her again although she was still in there. Somewhere. She grew, and in a few months the Forest Service brought her to the soon-to-open Alaska Sea-Life Center in Seward where much larger tanks were there for her. She became their first octopus, an ambassador of her kind to people curious about life in northern seas.

Ophelia was a chimera—her recognizable and empathetic gaze arose from prominent eyes on an amorphous, unfamiliar body, slimy and writhing with boneless appendages. Cordova was equally chimeric. New here, I sometimes occupied a world of familiar small town charms but more often navigated uncharted shoals. I had never lived on the coast; I was not comfortable on or near the water—I had no experience with boats, nor saltwater, nor fisheries, nor underwater animals. Years earlier I had taken a scuba course but trained only in fresh water and had not been underwater since. I found it terrifying to draw breath while submerged.

The bush town of Cordova, Alaska, adheres to a steep mountainside below Eyak Lake and above murky Orca Inlet. The lake drains beyond the mountains, where the plaited watery tresses of the Copper River fashion a delta almost fifty miles eastward. At the town's feet begins Prince William Sound, a deep and fingered embayment of protected waters sheltered by islands and ringed by the retreating glaciers of the Chugach Mountains. The sound extends a hundred miles to the west.

Moose are familiar wanderers through the yards of residents, shockingly tall next to cars and fences. Bald eagles perch on poles over the harbor. Flocks of blue herons from the roosting colony a few miles up the inlet sometimes occupy the trees on Spike Island, which I can see outside my office window, above visiting killer whales and wheeling mobs of Glaucous-winged Gulls and Arctic Terns that swoop and peck over the water. The town is not connected to an outside road system. Arrivals come by boat and ferry, by small plane or government-subsidized passenger jet service that stops at the Mudhole K. Smith airport on flights between Seattle and Anchorage.

Cordova Harbor is the briny heart of town, bursting with low-draft bowpickers. These thirty-two-foot aluminum boats are specialized for working seine nets over their bows in the delta shallows to pick the fatty red salmon of the Copper River. The salmon are at their peak before they enter fresh water on their long spawning migration across the flats and up the 290-mile length of the Copper. The clamming also was wonderful around Cordova until the 1964 Good Friday earthquake. Shifting blocks of the earth lifted miles of mud-flat clam beds above the high tides, and so those clam-rich times were now faded but fond memories.

Little encourages science to be interested in Alaska's octopuses. Unlike the salmon, halibut, and herring, there is no commercial fishery for octopuses, and so fishers regard the occasional catch with interest or cut the octopuses up for halibut bait. The cut arms sway enticingly in any surge along the bottom, and the tough lean flesh stays fresh on the hook for some time. Without a commercial fishery for octopus, there are no annual surveys and allocation of funds within a management agency; there is no one counting the octopuses, no one asking what they need to live and how they find it, no one funding researchers.

Alaska's fisheries alone support the careers of hundreds of fishermen and women. Fisheries biologists monitor, count, and work to understand the populations of salmon, halibut, Pacific cod, and crabs that enrich our waters and the state coffers and drive Alaska's industry. In 1995, I could find just three biologists studying these octopuses, each of them doing it on the side when their livelihood permitted. None lived or worked in Alaska.

Octopuses are harvested, however, in noncommercial ways. Among the Eyak and Sugpiaq and other native peoples who live along the southern Alaska shores, octopuses were both subsistence comfort food and breakwater against historically rising tides of cultural erosion. This accorded the octopus a place on the list of managed resources harmed by the oil spilled from the supertanker *Exxon Valdez* when it ran aground on Good Friday in March 1989. Some funding was therefore committed to better understand this harvested food item.

My training specialized in animal behavior. I had no experience in marine systems in Alaska, nor with native cultures. I was, however, available, eager to work, and living in the right place, with technical and scientific skills. I submitted a proposal to join with those harvesters. I hoped to collect new data and to learn from elders of the coastal communities. Mine was the only proposal submitted for octopus research; I was funded.

Cordova sits on the traditional lands of the Eyak people, of Dënéndeh (Athabascan) heritage. The last traditional Eyak village was annexed to the town in 1900. Cordova was the birthplace in 1918 of Marie Smith Jones, who by 1995 was the last surviving native speaker of the Eyak language.

Forty miles northwest of Cordova, at the base of Ellamar Mountain, lies Tatitlek, a Sugpiaq village of about a hundred souls. Tatitlek is the nearest village to the grounding location of the oil tanker *Exxon Valdez*. Millions of gallons of oil spilled in this disaster. Oil landed on beaches and damaged the health of marine animals, severely reducing the use of wild subsistence foods by these villagers, as well as by Alaska Natives along more than a thousand miles of shoreline touched by the spill. Sixty miles southwest of Tatitlek on Evans Island lies Chenega, another Sugpiaq village of less than a hundred people whose ancestral home in Old Chenega was destroyed by a tsunami after the 1964 Good Friday earthquake. After an exile of decades, many village residents relocated in 1982 to the new Chenega (also known as Chenega Bay), a few miles from the original village site. In these villages, Alaska Natives introduced me to some of the uncultivated places where they find their food. Giant Pacific octopuses live in the waters of Alaska. In good years, these are an important food item that's harvested in coastal villages. There is a lot of meat in such giants.

I asked in what kinds of habitats could I find octopuses. Those familiar with the glaciated shores and cold green waters of Prince William Sound said, "Octopuses are where you find them." They are everywhere in any marine habitat, and nowhere, being only rarely encountered. They told me that octopuses could not be studied.

This statement proved true in one sense: many established scientific techniques used in fisheries and behavioral ecology would not transfer well to octopuses. But it proved false in another sense: octopuses were hard to study, but maybe it was not impossible to do so. Octopuses are flexible, boneless, slippery, and elusive. They are also clever predators. Predators in general are relatively rare in nature. Their population numbers are not as large as the population numbers of the prey required to support them. This was familiar to me from my work as a predator biologist, watching African lions and wild dogs cross the Serengeti, or hoping for glimpses of wolves and bears in Yellowstone. The limited information, the tall tales, and the warnings from locals, even the stories and legends, stood as signposts, marking the edges of the unknown.

# 2

# Dangerous Giants

My proposal promised that I would find and capture octopuses in Prince William Sound, while scuba diving, wrestling the animal into some sort of bag or container. Once captured, they could be weighed, their sexes identified, and the species of each octopus assessed and documented. These are necessary data to understand the dynamics of an animal population. Parts of this plan would cause me concern, first among them the idea to capture octopuses—just how big would the octopuses be?

Giant Pacific octopuses, the species most often identified in the sound, grow to between thirty to sixty pounds at maturity, according to reports from Washington State. This is as much as a full-grown German shepherd.

Nevertheless, the oceanographer of the Prince William Sound Science Center in Cordova related to me a story she had heard: an octopus had drowned a medium-sized dog that was harassing it in shallow water. Gary Thomas, the Science Center's president, told me of hooking a ninety-five-pound octopus while fishing for halibut on the backside of Spike Island, right outside our door. Up on First Street at the Cordova Historical Museum, I learned the tale of a diver who got too close to a big octopus that grabbed him by one leg. The diver was on surface-supplied air and connected by intercom, so he was not in danger of running out of air. The octopus held him underwater for two or more hours before he was able to free himself. I was not entirely sure I should put too much credence in the story, but it still gave me pause. What sort of risk did I face in handling these wild animals?

I began arrangements to work in Tatitlek, Chenega, and Port Graham. Along the way, I heard more octopus stories from Alaska native history. These older local accounts were not reassuring, nor were those from other shores. Around the time when many old ways and native languages seemed like they were dying out in Alaska, anthropologists from Pennsylvania and from Denmark came to Prince William Sound to talk to the Alaska Natives there. Galushia Nelson shared stories from his wife, Annie. Nelson told them that there is a deep hole outside of North Island in Cordova Bay. A devilfish lived in the hole and grabbed canoes that came by at night. As the devilfish rose from the bottom, the water around the canoe filled with slime and those in the canoe could not paddle away. Then the arms of the devilfish came out of the water. So the Eyak used to warn people not to go there. A boiler from a fishing boat was thrown into the water. That drove the devilfish away. But Nelson had other stories of giant animals, as well. A giant bear, a beaver; these giants were called et'stli'yatl, and they lived in the ground or under the water. As Nelson said, they were all bad, and they ate people.

The Eyak tested themselves against the giants. Old Man Dude related how two brothers went out to hunt. They came to a place where the water grew murky and the canoe bogged down. Soon the brothers could make no progress at all. Beneath them, the water turned brown and a huge devilfish emerged from the gloomy depths, swarming up to meet them. Its legs were as thick as dishpans. The two brothers strapped knives to their wrists and leapt into the water to do battle with the giant. They cut and stabbed at the devilfish until one stabbed it in the heart. The devilfish died and floated up. It was one hundred and fifty yards wide; and where it floated on the water, no one could paddle across it for the water was too thick with its slime.

Stories about octopuses of this size are hard to take at face value. Yet

I found reports from other areas of octopuses of similar size. An 1897 account from the Smithsonian Institution reported a carcass, apparently an octopus, weighing several tons. There were too many stories simply to ignore.

I had no experience with octopuses nor other ocean animals. Stories recounted by old salts intimate with the sea raised questions or piqued my curiosity, but even this knowledge was limited by the inaccessibility of underwater realms. This was the first time, but would not be the last, that stories, whether anecdotal, scientific, or from Indigenous cultures, would stimulate me to follow trails and tendrils of ideas through old and recent publications, at times into my own field work. It was not until later, after years of working with octopuses across the Pacific, that I sometimes was able to rely on my own experiences for guidance and inspiration.

I paid closer attention to records of octopus size: Just how big does a giant Pacific octopus get? And how would I get into the water with them and work safely?

IN THE COURSE OF RESOLVING these questions, I found myself on a short whale-watching trip to the Santa Cruz Islands off California, seeking a glimpse of blue whales. Also on board was Clyde Roper, a Smithsonian researcher engaged in a search for Architeuthis, the giant squid. Blue whales are the largest whales—indeed the largest animals—ever to exist, and the search for whales and squid kept our attention handily focused on ocean giants.

Clyde was a muscular man at that indeterminate age past fifty where men are mature but not yet elderly. His trim gray beard and New England accent were marks of a man long at home on the sea. I asked Clyde for more about his search for Architeuthis.

"It's really very exciting," he said and proceeded to fill me in. Architeuthis is a true giant among animals: the largest invertebrate ever. It was believed by scientists at one time to be a mystical beast. Linnaeus dropped the giant squid from his book *Systema Naturae* after the first edition. However, no fewer than forty-four carcasses of Architeuthis washed ashore between 1853 and 1898, most of them in Newfoundland, which dispelled all doubt of its existence. The largest specimen was estimated to be fifty-six feet long when found in 1878 in Thimble Tickle, Newfoundland. The species once was thought to grow as long as fifty-nine feet and weigh 1,980 pounds, but current and more accurate maximum estimates of forty-two feet length and 500 kg (1,100 pounds) are impressive enough.

Beyond its size, we know little about the giant squid. At the time of our whale cruise, no one had ever seen one in its natural habitat. Enter Clyde Roper. By mapping where carcasses appeared over the decades, and over years spent accumulating clues to the squid's secret life, Clyde pinpointed likely areas where these squid dwell.

Clyde used one giant to search for another. Off the shores of New Zealand lies the Kaikoura Canyon, a marine gorge that within a mile of shore drops to a depth of a thousand meters. Sperm whales, another of nature's giants, frequent this canyon. These whales feed regularly on cephalopods, particularly the giant squid and their mid-water relatives. The cephalopods include squids and their evolutionary siblings: the cuttlefish, octopuses, chambered nautiloids, and now-extinct ammonoids and belemnites.

With a large suction cup, Clyde mounted a video camera on the head of a sperm whale. The camera stayed on for only thirty minutes to an hour, but in that time the whale took the camera down into the depths. Clyde could see what the whale saw. By putting these critter cams on sperm whales in Kaikoura Canyon, Clyde hoped to follow them to the giant squid, which were both their quarry and his.

As we talked, the captain of our whale-watching ship drew our attention to a pod of blue whales about a mile off. The boat turned and headed in their direction. With our attention fixed on the whale spouts in the distance, we did not notice the looming dark shadow beneath the bow of the boat. It grew until, with a sudden loud blast, a geyser of whale breath erupted from just beneath the surface and showered us with salt water. Looking down from the deck, Clyde and I stared into the open blowhole of the largest animal ever to live. A huge table of dark blue flesh rose below us, bumpy with barnacles and crossed with lined scars. The whale filled its lungs with a roar of wind and the blowhole closed as the whale sank again.

"Whoa! Whoa!" shouted the captain over the loudspeaker. "Did we hit it?"

We had not, but the whale was no more than inches from the boat. The whale seemed bigger than the boat; somehow larger than life, like bedrock, an animated piece of the earth itself.

EVEN AT THE HIGH ESTIMATE for Architeuthis of fifty-nine feet long (the length of an average blue whale) and weighing nearly a ton, science had debated the existence of even larger cephalopods than Architeuthis.

The Colossal Octopus entered the realm of science late in 1896 following a great storm, after which a carcass lay half buried in sand on the beach south of St. Augustine, Florida. The carcass was mutilated at one end, badly decayed, and very large. It measured twenty-one feet long, seven feet wide, and four and a half feet high. It likely weighed several tons. The main mass was sack-like, with numerous tendrils of flesh found around, and in some cases still attached to, the carcass. This lent it the appearance of an astonishing octopus. It was of unprecedented size for any octopus, weighing several times more than the largest giant squid and with an arm span of nearly two hundred feet. Debate about the nature of this carcass began at once and continued for the next century.

In St. Augustine, Dr. DeWitt Webb examined the carcass in detail. He forwarded his observations to Professor A. E. Verrill, a scholar at the Smithsonian Institution well known in his day for his many publications on the giant squid. The appearance of stumps (as of arms) at the mutilated end of the carcass led Verrill to publish a description of the animal. Verrill named it *Octopus giganteus*.

Verrill soon received from Webb specimens of tissue from the carcass. When he opened these, he realized he was wrong. Verrill had examined many carcasses of the giant squid, and he well knew that the bodies of cephalopods, even giant ones long dead, are composed of hardened muscular fibers. However, the tissue he received from Webb instead was an "elastic complex of connective fibers." As the tissues also "smell like rancid whale oil," Verrill concluded they were the remains of a whale. He published a retraction of his identification of the carcass as a Colossal Octopus.

Sea monsters, however, do not die such easy deaths. The Smithsonian Institution preserved the samples of tissue from the St. Augustine carcass. Seventy or more years later, Forest Wood, a founding member of the International Society of Cryptozoology (a word coined in the 1940s to describe scientific attention to lake monsters, sea serpents, Sasquatch, and other unverified wildlife) and then at Marineland of Florida, unearthed newspaper clippings about the carcass. Wood searched the events following its discovery. Upon learning that samples from the carcass still existed, Wood had them examined by a cell biologist at the University of Florida. Together they wrote an account of their explorations, concluding that the St. Augustine carcass "was in fact an octopus." The Society's *Journal of Cryptozoology* published an article relating the astonishing fact that, when examined by modern methods of histology, the tissue most resembled that of an octopus.

The carcass at St. Augustine is not the only one of its type. Similar carcasses have washed ashore in New Zealand, and as recently as 1988 one was found in Bermuda. Collectively they are known as globsters (one in Bermuda was dubbed the Bermuda Blob). A number of experts, including Forest Wood and Clyde Roper, examined samples from this latter carcass. The definitive analysis, however, came from a group of scientists who obtained a piece of the St. Augustine carcass, as well, and subjected both to the detailed scrutiny of electron microscopy and amino acid analyses. Their conclusion was that the remains were of vertebrates, and that "there is no evidence to support the existence of *Octopus giganteus.*"

In 1995, when a brief news article appeared in *Science* trumpeting the demise of a sea monster "under the probing gaze of a modern electron microscope," defenders of *Octopus giganteus* were moved to write a rebuttal. In brief, the rebuttal said that to witnesses at the time, the St. Augustine carcass looked like an octopus, and therefore we cannot dismiss the possibility that *that* is what it was.

Critics have argued that, whatever the St. Augustine carcass was, it could not have been whale collagen as Verrill ultimately decided that it was, as this does not normally exist in two- or three-ton lumps. However, investigations of similar monster carcasses, such as the Newfoundland Blob that washed ashore in 2001, using modern molecular methods, identified these carcasses as the remains of sperm and fin whales. These globsters were the remains of whales. This is the same conclusion Verrill had reached a century earlier using only his sense of smell.

How is this possible? Where are the bones and other recognizable anatomy such as organs or blubber? Whale blubber is quite tough and fibrous. It can be up to 30 percent protein, and half of that protein is collagen. A whale will leave behind tons of collagen, even after the oil in the blubber decays. A male sperm whale can reach fifty thousand kilograms, or roughly fifty-five tons, so that if just a quarter of a sperm whale is blubber, then two-ton lumps of collagen do in fact normally exist, and might well wash ashore on rare occasions.

As the dead whale drifts on the seas, the bones are heavier than the buoyant blubber. The bones tear out of the decaying flesh and sink. The recognizable organs and soft tissues decompose, leaving nothing but the inert collagen and the stench of rancid whale oil.

The spermaceti organ of a sperm whale, also collagenous, might account for the long attachment tendrils and sack-like shape of the St. Augustine carcass that first led Dr. Webb and Professor Verrill to the mistaken identification of *Octopus giganteus*.

---

WITH THE COLOSSAL OCTOPUS out of the picture, were there still reasons to fear the large size of the giant Pacific octopus? This was a matter of some worry to me when I started the study. I proposed that the divers capture these animals to obtain measurements. Was this plan unacceptably risky? Octopuses have twice as many limbs as my divers and possibly would outweigh them severalfold.

Our research team gathered for the first time in the Science Center conference room on a rainy afternoon in July of 1995. Outside the window, gulls hung like dust motes over the mouth of the harbor, while sea otters paddled lazily on the waves. My experience with marine research was limited to only a few months. Despite my lack of experience on and under the high seas, I was the expedition leader and accountable for the safety of everyone, the result of my successful proposal.

For the research, I chartered a wooden trawler that had been converted into a research vessel named *Tempest*. Our expedition would consist of a science crew of six, plus vessel captain, Neal Oppen, and a single deckhand. Among the science crew was my wife, Tania Vincent. Tania was officially along as a volunteer researcher—she did not want to stay at home while I went out and had all the fun. Although Tania shied away from the administrative duties, which I bore, she took on the responsibility for the design and execution of sampling plans.

Dan Logan and Michael Kyte were our divers, with Neal as an alternate diver. Dan dove with the US Forest Service; he worked with us for two years to keep his skills current. Michael had captured octopuses for the aquarium trade in Puget Sound for ten years or more, and was now a consulting marine biologist. I invited him to join our team when reviewers of the project suggested that some additional expertise might be helpful. Our other team members that year were two field assistants, Scott Wilbor and Kathleen Pollett.

Michael gave us a rundown on octopuses, and how to handle them underwater. He introduced us to octopus life history, familiar to me through my reading. He then briefed us on his methods and techniques for capturing octopuses.

"This is for the divers now. I'll be down there with you, so I want you both to know what to do at the right time." Checking that he had their attention, Michael looked first at Dan then at Neal. Michael went over the basics of how to identify a den, how to approach from above and behind (so that the diver did not stir up silt and the octopus did not get agitated), and how to scoop the octopus as it is moving out of the den, before it has a chance to grab hold of some rocks.

"Once it's got a hold of the bottom, you are not going to be able to pull it up without injuring it," he said, "so it's important to scoop it up free of the bottom quickly, and then handle it in mid-water."

Tania, Kathleen, and I were new divers, without the experience of Neal, Dan, and Michael. For a beginning diver, one of the most difficult tasks is to maintain neutral buoyancy so that you can work in the water column, neither popping to the surface like a cork nor dropping to the bottom in a dense cloud of silt. We novices glanced at one another, faintly intimidated at the thought of having to juggle an octopus and maintain neutral buoyancy at the same time.

"Once you've got it off the bottom," Michael continued, "then you can point it body-first into the bag and it will just glide right in."

"Now if we get a really big one," Michael paused, "I'll handle it. But everybody needs to know what to do in case there are problems. It can be more difficult to scoop a hundred pound octopus off the bottom than one that is only five or ten pounds." The trick with the larger octopuses, according to Michael, was to keep pushing them up and away from the bottom and turn them head over beak again and again. Eventually, feeling harassed, the octopus would adopt a defensive posture curling its arms over the body and forming into a ball, with all the suckers facing out. In such a pose, arms held close to the body, an octopus will not be holding on to much, and the diver can more easily scoop it into a bag. Now even the experienced divers seemed faintly impressed at the thought of juggling a hundred-pound octopus in mid-water.

"Of course, with the big ones," Michael went on, "you have to be a little concerned about letting the octopus get a hold of you. Even a medium-sized one is strong enough to pull your mask off or to take your regulator away, but you can't really get into trouble unless..."

(I fleetingly wondered here what Michael considered trouble, if not an octopus stealing a diver's mask and regulator.)

"You can't really get into trouble unless the octopus is larger, in which case if it gets wrapped around you, it can pin your arms to your sides and you won't be able to retrieve your mask.

"Of course, that's not really serious," Michael added, "but if it has taken your regulator while your arms are pinned..." I began to see where Michael got his sliding scale of severity for trouble, although drowning with your arms free did not seem that much preferable to drowning with them pinned.

"I'm sure we won't have any trouble," Michael said and looked around at the divers again. "But if we do, remember not to panic. Don't fight the octopus directly: collectively, its suckers are stronger than you are anyway and it is too slippery to hold on to. Instead, pull an arm off one sucker at a time, like you would pull a bath mat off the bottom of the tub. If you're stuck, just let go and relax. The octopus may also let go and retreat."

Dan asked about the risk of being bitten, and Michael explained that it was possible, but not a big risk if you were careful. "Don't put your hand on their mouth," he said.

When Michael talked about big octopuses, he meant somewhere in the range of seventy pounds or more, the size that he encountered while diving in Puget Sound. The way people talk, however, I wondered if we might not find even larger ones in Prince William Sound. The octopus record that no one disputed was 156 pounds, captured just north of Victoria, British Columbia, in 1967. It measured almost twenty-three feet, from arm tip to arm tip. I heard about this one from James Cosgrove, the chief of Natural History Collections at the Royal British Columbia Museum. He saw this octopus himself, on display for a few weeks at the Pacific Undersea Gardens until it died. Accounts of octopuses weighing about one hundred pounds are not that rare.

There are occasional reports of much larger animals, however. At the Santa Barbara Museum of Natural History, there is a photograph of a fisherman, Andrew Castagnola, with an octopus that he caught in the Santa Barbara Channel off Santa Cruz Island in 1945, for which the stated weight was 402 pounds. Another one caught off Santa Catalina Island early in the twentieth century appears in a second photograph in the same collection, and looks about the same size, although no one weighed it.

Jock MacLean's reports from the 1950s top all claims for the species. MacLean was a commercial diver and fisherman in the Johnstone Strait, along Vancouver Island. He captured one octopus weighing 437 pounds and described another that weighed about 600 pounds and measured thirty-two feet across. In one account of Jock's tales, the 437-pounder "filled a forty-five-gallon barrel" and was weighed or measured. This tale was a yarn told on a boat several years after the events. Jock never captured the 600-pound giant but sighted it and estimated its length. If this is the case, then these weights and measures are highly suspect. Stories grow in the telling and such estimations can easily be off by a factor of three or four. All of these weights are more than double the largest size of well-documented records.

However, in another version of the events, Jock captured several animals of around 400 pounds; and the 600-pound octopus was captured and weighed. In this version, the measurement of thirty-two feet was from "tip to top" (arm tip to top of mantle), not, as I found more often, from "tip to tip" (arm tip to arm tip, extended). That is, spread out on the seafloor, this individual could reach about ten yards from its mouth. In all, it would be about sixty feet wide. Like the Eyak monster that ate people, no one could paddle past this one, at least not in a stroke or two.

Records for the species of 400 pounds and more are from over half a century ago. If such giants ever did exist, it is likely that they are now of the past, like giants to be found, once upon a time, in fairy tales. It is ironic that if such animals do still exist (assuming that they ever did), they are not to be feared, nor battled and subdued. We have gone so far by way of penetrating the last wild places of the Earth that now even our half-mythical beasts and monsters are in need of protection. The giant or even the Colossal Octopus can no longer be the personification of the Terror of the Deep. It would be a strange and sad fate indeed if we managed in our exploitation of the Earth to deprive ourselves not only of its continually renewing resources but also of our myths themselves.

As for me, I will someday scuba dive into the waters around Vancouver Island, looking for legends from past ages. Of all the places I have read about, Puget Sound seems to have the largest and most abundant octopuses. While I am down there, swimming through the filtered half-light of the underwater world and listening to the slow, deliberate rhythm of my own breathing, I hope to see a truly large octopus out on his daily rounds. When I do, just once, I won't try to capture it and take it to the surface to weigh and measure. I will find some way underwater to estimate its size. Sure, I may be off by a factor of three or four in knowing its weight, and some critics will claim that it was not as big as I thought. But I won't mind; they won't have been down there with me, immersed in the story. They won't know what it really is like to glide through the water with half-mythical giants.

# 3

......

# Lost Homes

My early research would be while scuba diving: counting the octopuses and surveying their habitats. But I also surveyed them on foot during very low tides, in their intertidal habitats. Both surveys depended on my ability to locate an animal notorious for its ability to pass unnoticed. Finding octopuses was no less a problem than capturing and weighing them.

To learn how to do this, I accompanied Alaska Native elders harvesting octopuses along the shore. Alaska subsistence harvest of octopuses traditionally occurs in the intertidal and on foot, during the lowest low tides early in the summer and midwinter. I was not yet a diver with enough experience to conduct the underwater surveys myself, but I could do the intertidal work. The start of my fieldwork in 1995 was to travel with Scott Wilbor as my assistant to the villages of Tatitlek and Chenega. At the time, I could find no published accounts of octopuses using intertidal habitats, although their existence there was common knowledge in Alaska Native communities. I have since discovered a few published reports that were available then of other octopus species in the intertidal, and more have appeared over the intervening years since.

In Tatitlek, we were to learn from Jerry Totemoff, a serious Sugpiaq man of forty-five or so with short dark hair, whom we met briefly on our first afternoon in the village. Jerry enjoyed the traditional subsistence foods, and he was the one who most regularly brought octopuses back to the village. I hoped that he would show us how he found octopuses, and the types of areas where he regularly located them.

Jerry arrived soon after we got to the Village Council office, dressed in a worn gray jacket over blue jeans and XtraTuf rubber boots, the standard footgear in this part of the world. To get to the sites he wanted to show us, Jerry suggested we leave only half an hour before the low tide the next morning. But I wanted to go earlier; I didn't want to miss our narrow opportunity to survey areas that were submerged for all but the very lowest spring tides.

"Them reefs won't be up till later," Jerry said. And so we agreed to meet only a half hour before low tide. Jerry shook his head at visitors who wanted to get out of bed early. I worried that we would get to the reef too late to see anything before the tides turned.

Jerry left, and Scott and I took our gear back to the Council office where we stayed. In the back room under a small gray window were two bare beds with thin mattresses. The front room contained a small white table, two chairs, and a hot plate. We used a public restroom down the hall, which lacked shower, bath, and hot water. We hoped not to have to spend too much time in the Council office. We went down to the beach to look at intertidal animals.

We knew no one in the village. When the Council office shut down at 4:30 p.m. we were alone in the building for the evening, with our one-pan hot plate meal of canned chili. We worked into the night trying to identify a handful of starfish, chiton, and tiny crabs with a technical and obscure guide to the identification of ocean animals.

We met Jerry at 6:30 a.m. the next morning. It was warm and perfectly still. A light rain fell on the quiet water of Tatitlek Narrows. As we approached the dock, a man called Old Eddie leaned out of the open window of a small shack, from which blue paint was peeling to reveal the gray wood beneath. He yelled, "Bring me back a octopus. I'll make soup!" Jerry lifted his hand in greeting and laughed, but he didn't pause, and we made our way down the steep bank to the water.

True to his word, Jerry had timed our arrival so that the reefs we visited were only just emerging as the tide receded. The first was an outcrop of rocks a few hundred yards down a gravel beach from the Tatitlek dock. Along the way, Jerry stopped at some alder bushes and cut off a straight section of green branch, from which he stripped all the leaves and smaller branches. This left him with a five feet long section that tapered from one end the width of a man's thumb, to a tip the thickness of a pencil. On reaching the rocks, Jerry walked over them to the seaward side and thrust his alder branch into a hole in the rocks, questing for an octopus. I asked Jerry whether he had found octopuses in this spot before.

"There used to be a lot of octopuses here, two or three every tide, but no more. Look here." He pointed to a hole in the rock, in front of which was a small spray of gravel. Several small shells scattered over this: a broken clamshell, the carapaces of two helmet crabs, a pearly section of rock oyster shell. To my untutored eye, it was not a remarkable collection of debris from the beach.

"Nobody home," Jerry said. It wasn't until this remark that I recognized the hole for what it was: the entrance to an octopus den; the first one Scott or I had seen.

Just as Scott and I needed lodging for the night, so octopuses need shelter. In Prince William Sound, circumstances and Alaska Native traditional knowledge allowed me to use octopuses' need for shelter as a way to track them. The technique for finding an octopus with a green alder branch is simple. If the bottom of the hole feels hard, you are poking bare rock—the hole is empty. If the bottom is soft, on the other hand, there is something in there. But what? If it's a starfish, or mud, or almost anything else, then you can poke and prod and not much changes. But when you poke an octopus in its den, the indignant beast usually reaches out an arm to grab the intruder.

"What is this?" she wants to know.

When you gently try to remove the stick from the hole, she may hold on or pull it farther in, leading to a tug-of-war over who gets to keep the stick. On one later occasion, I had an octopus take the stick entirely into her den where I could no longer reach it. I waited. The octopus usually tires of the stick when she decides she can't eat it, and after pulling the stick into her den, she will then push it back. And often, having decided (correctly!) that someone is after her, she will retreat farther into the den, and the second time you probe the hole, the stick will find only bare rock at the bottom.

Octopuses are finicky about where they live. The shores of the sound and its underwater slopes are littered with rocks, piled on top of sand or mud or more rocks, jutting up out of the substrate, forming vast walls of bedrock or boulder fields hundreds of yards wide. Yet most of these rocks are not suitable for octopuses. I seldom discovered octopuses denning in boulders that jutted way down into the sediment or under boulders that rested on bedrock or other big boulders. Instead, octopuses most often choose boulders resting on mud, sand, or gravel, any of which they can readily dig into to make a cavity beneath the boulder. The excavated materials form a berm around the opening to the den.

Having found or constructed a suitable den, octopuses tend to return home each day to the same hole. If the resident moves out, or is killed, another octopus may soon move in, so that over the years and decades the same holes turn up occupied again and again.

Like a trapper working his lines, Jerry periodically checks the best dens in a stretch of beach and collects the residents. In a few days or weeks, new animals will have moved into the vacant homes.

Octopuses build a back door or a minor entrance that looks less used than the front door; and the intertidal dens we found almost always contained a pool of water, so that although both doors may have been above water at low tide, the occupant sat curled up snug inside his or her own little homemade tide pool. These dens are wonderful hiding places. Only their midden pile—the berm of pushed up gravel, a few scattered shells—and a small opening mark the octopus's home. When tides submerge the dens, the underwater residents sit outside their dens, on the porch, as it were, and watch the world go by, perhaps waiting to see crabs or other prey creeping through the kelp or eel grass lawns just outside.

In need of peace or safety, the octopus can slip inside, and only the slenderest eel-like predator can follow. Without hard shells, octopuses fall easy prey to a number of predators, especially bottom feeders like sharks, flatfish, seals, or sea otters. A surprise attack by a larger shark or otter can quickly overwhelm an octopus, but their den is a safe refuge. Once inside, an octopus is out of reach of almost any predator. Even for humans, capturing an octopus from inside its shelter is not easy. Jerry told us how to extract octopuses by spearing them through the head with a hook to pull the remains from the den.

Fortunately, for the octopuses, we were studying them, not harvesting them. The outcrop where Jerry brought us was the uppermost section of a jutting layer of tilted turbidite. This rock originates when silt, gravel, and sand collapse down the continental shelf in a great underwater avalanche of mud (called a turbidity flow) that spreads out on the seafloor. This material is eventually buried and transformed to rock. On this beach, through the slow movement of continental plates, the turbidite was brought to the surface and exposed. Uplift tilted the rock off the plane of its formation, so that layers running through it pointed up into the air on the seaward side, and down into the earth toward shore. Lush beds of eelgrass lay on either side, just covered by the water.

Large purple sea stars clung to the sides of the rocks and opalescent nudibranchs, a marine version of the garden slug, crawled about in the small tidal pools. The nudibranchs slowly waved their four front appendages, each edged with pearly blue. Rows of brick-orange cerata glistened on their backs while they grazed the hydroids and other attached animals that grew on the rock. Neither Scott nor I knew these animals well enough yet to give them their Latin names, but we would learn these and many more over the next few weeks as we struggled to come to grips with the octopus's environment.

In some sections of the outcrop, parts had crumbled, leaving fissures and holes that ran deep into the rock. It was into these fissures that Jerry probed with his stick, rapidly moving from one hole to the next. I followed him, trying to record the details of each lair he examined, so we would have data on what kinds of areas these animals were using and those they were not. I was sticky with heat inside my rain gear. The stillness of the air was stifling. Fortunately, there were few bugs as Scott and I moved along behind Jerry, watching.

"Nobody home," Jerry said, which I took to mean that an octopus had left this debris here, but was not in the hole now. I looked on with more interest, and called Scott to come over. We were counting on Jerry to teach us how to recognize the track of the octopus; and he was counting on our scientific training to learn more about the state of the animal across the entire sound. Jerry quickly moved on to other holes. Since they, too, were empty, he went back to the base of the outcrop, ready to take us by boat to our next stop.

We arrived at the barely visible tip of a similar rock outcrop, located in several feet of water just off the Tatitlek dock. There were fewer holes here, but at one, which in all respects appeared unremarkable, Jerry said "I think somebody's home." We had found an octopus. It weighed three and a quarter kilograms (about seven pounds), and measured about half a meter long (a bit shy of two feet). When we found three more octopuses at another site later that day, my worries that there would not be enough octopuses for me to study vanished.

By ten o'clock that morning, the tide had risen, the reefs were submerged, and we were headed back into the village. Jerry dropped us off and disappeared, leaving Scott and me to our specimens and our books for the remainder of the day. We had collected shells from in front of each octopus den, to learn what the octopuses were eating. Between the rain, our invertebrate studies, and our status as strangers in the village, Scott and I spent most of the next three days in our rooms, leaving each morning for the beach work but for little else. By the time our plane arrived in the afternoon of the third day, we were glad to be on our way.

"Where are we staying in Chenega?" Scott asked me as we loaded our gear into the plane on the Tatitlek runway for our flight across the sound. A stiff breeze warned me that takeoff would be bumpy. I handed the pilot a pair of hip waders, which were useful for our beach landings, and a duffel bag. He stowed them in the rear of the plane.

"A few weeks ago, Mike Eleshansky told me we could stay in the schoolhouse." I shrugged. "I haven't been able to reach him since, to check whether he remembers we're coming." Each village had a designated facilitator, whom all researchers were to contact if they wished to work in the village. Like many Sugpiaq, Mike was very cordial. But it was hard for me to read his tone or manner. Our cultures were different enough that at first I could not tell whether he was intense, or joking, or bored with me, or if he found me rude or tedious or engaging.

I had asked Mike about lodging, and he said that we could probably have a spot in the schoolhouse—he wasn't sure though. Scott was worried, but to my mind, we could endure our lodging, whatever it turned out to be, for four days, as long as we were out of the weather.

Thirty minutes later, the plane touched down onto calm waters, and we taxied into the harbor of the village of Chenega. We found Mike at the Village Council office. Mike was a compact man of about sixty, with Sugpiaq features, close-cropped hair, and a prominent chin. I introduced myself, and reminded him of his plans to find us space in the school.

"No, them guys are in there. Let me think," he said. The space in the schoolhouse was full up—the whole town was, because Chenega was getting a new dock put in and all the lodging in the village, which wasn't much, was full of construction workers. Mike put his chin in his hand and thought, mumbling, "Let me think" every few seconds.

Finally, Mike looked up. "Nope," he said. "Everything is full."

After a little bit, Mike said, "Come inside. I think better in there," and so I trailed him into his office. Mike sat down behind a desk, put his chin in his hand, and muttered, "Let me think" again. There were two desks in the office, where to all appearances they had spent the last twenty years undisturbed under their mounds of papers and reports. A straight-backed chair, where another stack of government forms was enthroned, occupied what little space remained. I pushed the stack back and claimed a small corner of the seat.

Mike had an idea and picked up the phone. He made a quick call, and then said to me, "You'll stay with us."

Suitable shelter put a new light on the remainder of our trip. Mike was a kind and engaging host. His house was about four doors down from the Council office, a small square prefab where he quickly made us comfortable. Scott and I slept on cots placed in the pantry, next to Mike's freezer and beneath neat rows of canned vegetables, packaged foods, and bottles of Tang orange drink mix.

That evening, as we discussed where to search for octopuses, Mike gradually turned more talkative and more at ease. I explained my hopes that during our visit he could show us where to find octopuses. He seemed to appreciate our interest and our reliance on his knowledge to find good sites for our study. Mike was keen to go to Old Chenega if the weather was good. It was two hours away in his skiff, and he had not been out there recently. Old Chenega was the original site of Chenega village, destroyed by a tsunami in the 1964 earthquake. About twenty years later, the village was rebuilt on another island twelve miles from the old site. As Scott and I talked with Mike about traditional times and areas for collecting octopus, I sensed that he was glad to have someone take an interest in it again.

"The kids today," Mike said, "don't go out much."

In his younger days when he lived in the old village, he hunted seals and other game.

"My blood boiled for the hunt. I would come back for gas for the skiff, and then go right out again."

Before now, I had not heard about the earthquake from someone who had been through it. The quake hit at 5:30 p.m. on March 27 in 1964, reaching a magnitude of 9.6 and releasing almost twice the violent energy of the 1906 quake that nearly destroyed San Francisco. In areas around the sound, the ground rose and fell in undulating waves three to four feet high, creating rough seas of land. After the quake, great waves from the south swept away the village of Chenega. Every survivor lost family to the water that day. Twenty-six people drowned from a village of about 120—meaning that over a fifth of the population was lost. Survivors spent the night and the next day waiting for help huddled and cold on a hillside above the wet slopes where their homes had once stood.

The next morning we left the house at 6 a.m., to make the skiff ride to Old Chenega in time for the low tides. For most of the calm and sunny trip across Knight Island Passage, we sat without talking, keeping company with our own thoughts and the water and the sunshine. Mike stood at the tiller, squinting into the glare off the water, guiding us to his childhood home. As we neared Chenega cove, he pointed out the old schoolhouse, a weathered white building leaning into the hillside, as though the old building and the ancient island each held the other up during the recent years, abandoned to their fates.

As we pulled in close to shore, Mike eased the throttle back and pointed to absent landmarks: where the dock had once stood, where the church had been. Then we beached the skiff and walked along the shoreline. Periodically Mike would point out a prominent boulder, usually high up on the shoreline.

"We used to get 'em under that one."

We climbed up the beach to the rocks he indicated, but it was apparent that each was too far from the water. They were right along the storm line, marked by debris thrown high onto the beach by winter storms that coincided with high tides. These clearly had not been regularly submerged since the earthquake. The 1964 earthquake had lifted large portions of the western sound. The land where we stood was nearly two meters higher and seventeen meters (fifty-five feet) to the south of where it had lain when Mike lived here.

Mike realized with surprise that most of the good octopus dens he knew in the lower intertidal zone from his days in Old Chenega were now too high on the beach for octopuses. Like his own home, the octopus houses had been destroyed during the earthquake.

In the lower intertidal zone, where forty years earlier Mike had collected octopuses from their dens, suitable boulders for octopuses were surprisingly scarce. I expected that more boulders, once well below the lower tides, would have been lifted by the earthquake to replace the ones that were missing, but this did not seem to be the case at Old Chenega. It may be that boulders accumulate at the base of cliffs, but rarely roll very far from where they are deposited. The cliffs of Chenega Island itself were the source for the many intertidal boulders where Mike had found octopuses in his youth. The quake may have lifted hundreds of years' worth of boulder deposits up, out of reach of the octopuses.

On that fateful Good Friday more than three decades past, the quake struck with no warning. In my mind's eye, I saw the people of the village, on a quiet and holy March afternoon, winding up their business of the day, gathering on the hillside to chat or in their kitchens for the evening meal. The peace was shattered by the distressed earth, which groaned and heaved and surged beneath their feet for several long minutes. Shelves tumbled off the wall in pantries, dumping cans and jars to the floor. In dining rooms, pictures of the Last Supper fell from the walls, splintering. Floors cracked. People were thrown from their feet. Buildings and trees swayed as though bent before a hurricane, and a loud roaring filled the air even though there was little wind. As suddenly as it moved, the island settled again in its new position on the seabed. A brief quiet descended over the village.

Ten miles from Old Chenega, however, across an icy strait of water and up a long narrow fjord, a gravel sill beneath a tidewater glacier collapsed in the tumult, causing a huge underwater landslide. Lacking its customary support, the glacier itself disintegrated along its face, sending unnumbered tons of ice from a hundred feet or more over the sea in a cataclysmic plunge into the water. The narrow fjord funneled the resulting waves across the passage toward Chenega Island.

On the Chenega shore, the tide was already dropping. Waters receded with deadly silence to an unnatural low in the minutes following the quake. Children playing along the shore first noticed the empty seabed, newly emerged from its ocean shroud to the thin air that filled with a briny reek of low tide. Lank seaweeds, deprived of their watery support, lay weak and dying on the miry, sodden new land. Parents or uncles were already racing down the hill to call the children to higher ground, knowing that waves were likely to follow the quake. As the children turned toward their parents' calls, the waters of the sound returned, racing up the beach with a deafening roar, tumbling boulders the size of houses as they came.

The first wave arrived as a rapidly rising tide. Surging up the island slopes, it pulled children from the hands of their fleeing parents as it swept ashore. Older children helping younger ones were lost in the swirling waters, which climbed higher and higher as the wave submerged the village. Then the second wave arrived, traveling fast and one hundred feet tall. It swept up boats, piers, and buildings, ripped trees from their hold on the land, and plundered life as it receded. The second wave smashed the village to ruins and swept the wreckage into the bay, filling it with broken houses and swamped boats. Yet a third wave rose and flooded the village again, after which the floating ruins had vanished without a trace. Survivors were afraid to use the schoolhouse, their only surviving shelter, fearing more waves, and so gathered higher up the hill beyond the reach of the water.

Like the villagers, the octopuses, too, had lost their shelters. As this small band of cold and grieving villagers huddled on a hilltop, gazing in the cold light of the full moon at the ruins below them, dozens of octopuses, maybe more, of all sizes crouched along the shore. Perhaps many of them were adapted to the spring low tides, which occasionally brought their favorite dens up above the waterline. They might have instinctively waited for the water to return in a few short hours. Others were left in the open, their boulders tumbled and dens destroyed like the houses of Old Chenega. But this time the receded water did not return. Octopuses too timid to move slowly dried out and died of asphyxiation in the open, or froze as the shallow pool of seawater in their transformed landscape drained away or turned to ice. Others crawled down the beach in the moonlight. But even these might not have fared well. Upon reaching the water, instead of the near-shore boulder field where they were used to finding many safe hiding places, they found a broad plain of silt and gravel where boulders were scarce.

We never did find an octopus that day on the shores of Old Chenega, where the octopuses' old houses clung near the tree line and few new ones were to be found where the salt water lapped the shore.

I thought about those octopuses back in 1964, and I hoped that, like Mike Eleshansky, they all arrived safe and sound at some haven, at some new place to call home.

Homes, the dens of octopuses, were the easiest way to find and count these elusive beasts. My key into their world would be through their front doors.

# *Finding Octopuses*

# 4

················

# Our Cousin Octopuses

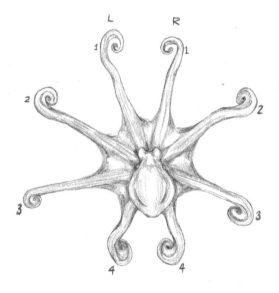

**Prince William Sound, Alaska**

I knelt on the rocks in hip waders. In a small tide pool before me was a tiny octopus, one of the smallest that my shore team and I had found. Male octopuses lack suckers at the tip of the third right arm, but on this octopus, on that arm, two rows of tiny suckers continued all the way to the end. It was a female. From tip to tip, she was about as large as my hand, and she weighed only seventy grams, a little more than an extra-large egg.

This was our first expedition with divers. The dive team was prepared to encounter giants underwater. But this was the second tiny octopus the shore team had found on our beach walks. It was the smallest octopuses that gave us trouble. Could these little creatures really be the giant Pacific octopus we were expecting? Or were they other species entirely? Grappling with this puzzle dragged me in far-flung directions to decipher the biology of octopus skin.

For the moment, the beach at Shelter Bay was deserted. The only sound to break the silence that settled over the bay was the steady drip of falling rain. Around the bend ahead were Scott, Tania, and Dan. Of the shore crew, only Scott had been on this beach before. He worked on seabird surveys following the *Exxon Valdez* oil spill. The bay faces north. The oil slick, carried southward by winds and currents, had poured in, coating and smothering the shorelines. Now, six years after the spill, waves and weather had scrubbed clean the hard, rocky beaches of Shelter Bay. Scott was visibly relieved that although oil still seeped from beaches in more than a few places around Prince William Sound, we did not see any sign of the spill here. He had dreaded returning here, recalling the awful damage.

There were few suitable homes for octopuses. This was not the type of spot where we would come to expect them. Hard rock outcrops formed a point on the east shore of Shelter Bay, where I now stood. In a small cove just to the east of the bay, the outcrops gave way to medium-sized cobble. We had just returned from the far side—three sea otters swam there and we had found lying on the cobble the chewed-up carcass of an octopus they may have lost. Our shore search this morning went quickly. We checked the crevices in the outcrops and looked for tide pools. There were no boulders to search among, although these were common on many other local shorelines.

We almost missed her. Only Dan's sharp eyes caught sight of a small carapace of the Oregon rock crab (*Glebocarcinus oregonensis*) at her front door. The bit of shell lay in a tide pool no bigger than a kitchen sink, on a flat shelf of rock that broke the steep descent of the beach to the water. A thin crevice ran through the rock above this pool. In the dark crack was our little octopus. After extracting her and taking her measurements, we put her back in her pool. I propped up a rock to give her haven from the sea gulls until the tide came up again to cover her.

Half an hour later, on our way back to *Tempest*, I stopped by the pool to see how she was faring. I tilted up the covering rock to peek, and there she sat, still looking exhausted from her ordeal with the biologists. Her arms curled in delicate spirals by her side, and her eyes were bright. She did not react to my presence. Unlike other octopuses we released, she had not yet crawled back into her crevice, which was still above water.

Muted reds, tans, browns, and cream colors spread slowly across her head and mantle, rolling in and out of one another. Waves of color washed down one arm to pass out the tip and then reappeared above to sweep down again. This was my first sight of the dramatic color changing abilities of cephalopods. The speed and flow of colors reminded me of the Alaskan northern lights.

DUE TO THE CHANGING NATURE of their appearance, it is not simple to recognize each different octopus species. That changeability makes it surprising that skin patterns are useful to identify species, but that is so, particularly for live animals. The identifying characteristics, however, are fluid and may not always be visible. The available field guides to marine animals were of little use to me at the time, given my inexperience.

More than three hundred different species of octopuses live along the world's ocean margins. Scientists continue to discover new ones. Still more species inhabit deeper waters. Among the coastal octopuses, females tend eggs until they hatch. There is no parental care after hatching. Some species, including the giant Pacific octopus, hatch into a planktonic life stage as paralarvae. After some time in the plankton, paralarvae shift their behavior from swimming to clinging. As they encounter and cling to debris or anchored algae, they transition to the bottom as juveniles and dwell on the seafloor the remainder of their lives. We have found these as small as 1.5 grams weight (about the weight of half a small grape). At this size, they hide in shells, kelp fasts, and under rocks on the bottom. Other octopus species have larger eggs and live on the bottom immediately after hatching. These coastal octopuses are solitary in nature, although there are exceptions.

Coastal bottom-dwelling octopuses have some characteristics in common. They have muscular, somewhat-spherical bodies and lack fins. At the center of the animal is the head with prominent eyes highly placed, and a downward facing mouth. Inside the mouth is a beak, somewhat resembling a parrot's beak. Surrounding the mouth are eight arms—a ring of prehensile lips—each with one or two rows of suckers on their undersides. A web of skin connects the arms, allowing the octopus to envelop the area between the arms almost as in a purse seine that is drawn into a bag to enclose the catch, or to flare the web between extended arms so as to appear much larger than it otherwise would. Behind and almost continuous with the head is the mantle, a muscular sac containing the body organs. The mantle cavity opens, and seawater is drawn in through gill slits, and is then ventilated out through a tubular funnel, the siphon. The gill slits make up a wide opening into the mantle cavity continuing under the head and body; and the siphon can appear on either side, emerging from the gill slit or underneath the animal. Most species have ink sacs and the octopus can eject an ink cloud from the siphon when pursued.

Two important features identified living giant Pacific octopuses, the species *Enteroctopus dofleini*, in 1995. In the technical taxonomic key that Scott and I had been using in Tatitlek the month before, giant Pacific octopuses are distinguished from other species by their large size and wrinkles or "extensive skin folds" on the body. What were we to make of these tiny octopuses? Even giants begin as babies, but perhaps small animals belong to another much smaller species—the red octopus (*Octopus rubescens*). The red octopus was sometimes identified this far north, but these reports, I much later surmised, were mistaken. This little species weighs less than one pound. If we were finding very young octopuses, size would not help us identify them.

We focused on the second trait. Octopus skin is textured and colored, in patterns that change but also are distinctive to species. In other aspects, octopus species can be very similar one to another.

An octopus has a lot of skin. Take a giant Pacific octopus out of the water and skin droops off its body like baggy, loose jeans. Put the octopus back in the water and for a second those extra drapes of skin float around its arms and body like silk scarves. Then a remarkable transformation takes place. The skin tightens up and takes shape. Two horns sprout above the octopus's eyes, bringing to life the apparition that earned them the old name of devilfish. Folds appear in ridges running from the head down the length of the mantle. A torpedolike point may form at the rounded end. The mantle folds break up into independent raised points—the papillae. The arms curl lazily in on themselves, forming neat little spirals that tuck against the web.

Glance away for just a heartbeat, then look back, and suddenly the octopus is lost against the background. If the octopus is in a patch of shotgun kelp, it may disappear entirely. Its reddish-brown skin matches the color of the kelp; the folds of loose skin mimic the rippling kelp fronds and sway in the ocean surge with the same motion as the kelp. No clear outline of an octopus remains.

Which of these diverse appearances were uniquely specific to the giant Pacific octopus, and which might occur among other species? The nascent internet at that time, then emerging as the World Wide Web, did not yet have the many color photographs of octopuses that are today available (albeit, even today, often with the species incorrectly identified). Our best key featured drawings not photographs. Even taxonomists were of little help. Taxonomic work until then was primarily with preserved octopus specimens, allowing voucher specimens to be preserved in museum collections where the same characters could be examined by later researchers. However, the distinctive appearance of the skin patterns in life is often absent in preserved specimens.

The red octopus was described as having skin with "small, pointed papillae," compared to "large truncate papillae" for giant Pacific. These descriptions of the skin protuberances of tiny octopuses felt hopelessly vague. How "pointed" was too pointy to be "truncate"? Good photos or drawings were scarce. For example, I can now clearly tell, although I did not know at the time, that the drawing provided in our key for the giant Pacific octopus was mislabeled—it included a false eye spot not found on this species, and lacked those characteristic mantle folds. However inadequate these descriptions, they did focus my attention on the skin as an important feature to note. It would take further research before I was able to determine that this small young animal in the tide pool was indeed a small giant Pacific octopus.

Skin has also been noted as a characteristic feature of octopuses by Alaska Natives, albeit this knowledge was embedded in their language. I discovered this months later, during a discussion about linguistics. Tania and I were at the Alaska Native Language Center at the University of Alaska Fairbanks. There, we spoke with two linguists, Michael Krauss and Jeff Leer. We had come out of curiosity. We knew only that Alaska Native stories about octopuses were in several languages, using different words for *octopus*, and that the Alaska Native words for octopus often were similar but not identical. I was curious how the languages were related, and what were the linguistic roots of the similarities.

Michael wore a frayed shirt and faded pants. He met us in the Language Center's library, a room so full of paper that it looked like a whirlwind had just passed through. The walls were lined with shelves, the shelves covered in books, the books buried beneath mounds of papers and manuscripts, and on top of the shelves, boxes piled to the ceiling, many bursting at the seams where documents threatened to tumble to the floor. Amidst this library of the world's written expertise on Alaska Native languages, Michael explained the history of Native words for octopus. Michael was an expert in the Eyak language, the near-extinct tongue once spoken on the Copper River Delta of Southcentral Alaska.

In Eyak, he told us, the word for octopus is *tse-le:x-guh*. Broken into its constituent parts, it means literally "rock under many-dwell" or "many things under a rock." Since octopuses in that part of Alaska can often be found secluded under rocks at low tides, *tse-le:x-guh* is a wonderfully descriptive word, as the eight arms of a single octopus might properly be regarded as many things. The Eyaks are related to the interior-dwelling Dënéndeh (Athabascan) people. It is not surprising that when they settled on the coast, they lacked a word for the octopus and created one by describing it. In the same way, the English language arrived at words for sea lion (an ocean animal that roars like a lion) and octopus (*octo* for eight and *pus* for foot, both from the Greek).

Jeff Leer, a linguist and expert on the diffusion of languages along the south Alaskan coast, joined the conversation. His disheveled clothes and grinning enthusiasm created the impression that perhaps he finds his work deciphering native languages almost too exciting. As he talked, he wrote the Alutiiq and Aleut words for octopus on a chalkboard. Michael pulled out a map.

Imagine Alaska as a fist made with the right hand, thumb and forefinger extended and the whole hand turned inward, palm down, so that the thumb and forefinger point toward the chest. The forefinger pointing to the left forms the Alaska Peninsula and Aleutian Islands chain, and the thumb pointing to the chest is the southeast Alaskan panhandle. The Eyaks on the Copper River Delta (from Cordova eastward) are at the base of the thumb, on the web between thumb and forefinger. The Sugpiaq (Alutiiq) live to the west of the Eyaks, and around the knuckle of the forefinger, which includes Prince William Sound, parts of the Kenai and Alaska Peninsulas, and Kodiak Island. The Sugpiaq word for octopus is *amikuq*; it is not descriptive in the sense that *sea lion* is in English, but stands alone, like the English words *horse* or *cow*.

The Aleuts (out along the extended finger) live farther west still in the Aleutian Islands chain. They have three words for octopus: *amĝux̂*, *aaqanax̂*, and *ilgaaĝux̂*, used in different parts of the chain. In between the territory of the Sugpiaq and that of the Aleuts lies Cook Inlet (at the base of the finger), an area occupied by the Dena'ina, who, like the Eyak, are of Dënéndeh (Athabascan) heritage, with ties to interior-dwelling people. The Dena'ina words for octopus are *amuguk*, *amiguk*, and *amaguk*, which are clearly related to the Sugpiaq's *amikuq*, and not to the Aleut words.

Over millennia, like paralarval octopuses dispersing in the ocean currents, the travels of Alaska Natives extended the roots of language wherever they went. The Sugpiaq are a coastal people. *Amikuk* appears in the language of other coastal aboriginal peoples of the circumpolar regions, from Russia (Siberian Yup'ik) to Alaska (Yup'ik and Inupiaq) to Canada (Inuvialuktun, Inuktitut, and others) to Greenland (Kalaallisut). *Amikuk* is like the Sugpiaq and Dena'ina *amikuq, amiguk* series. Just west of Cook Inlet, over the knuckle and on the north side of the Alaska Peninsula, lies Bristol Bay and the coastal territories of the Central Yup'ik. These words clearly all have a common ancestry, coming from the Inuit-Yupik-Unangax̂ language family.

Yet *amikuk* in Central Yup'ik does *not* mean octopus. The octopus of the southern Alaska coast is the giant Pacific octopus, a species that can top one hundred pounds and is the stuff of legends, Hollywood monster films, and Jacques Cousteau high-seas drama. The Central Yup'ik coastline, however, lies almost entirely outside of the range of this species. The giant Pacific octopus is found from California to southern Alaska and over to the waters of Japan and Korea, but its range in the Bering Sea seems limited to the shelf edge from around Unalaska to approximately Cape Navarin, Russia, and the range does not get much farther north in the Bering Sea nor close to shore along western Alaska.

So, what does *amikuk* mean in Central Yup'ik? The word names a legendary octopus-like creature, described as a sea otter without fur. The Central Yup'ik *amikuk* of legend is impossible to capture, and when killed multiplies Hydra-like: eight creatures arising where one was lost. The set of related words among Central Yup'ik, Sugpiaq, and Dena'ina surely results from the close interactions of these three peoples around Cook Inlet. Michael and Jeff speculated that the Central Yup'ik at one time occupied the Cook Inlet area. From there, the Dena'ina may have driven down into their range, splitting off a group to the east and south that became the Sugpiaq, and driving the westerly Central Yup'ik off the shores and out of the range of the giant Pacific octopus. Without any live octopus to relate it to, the *amikuk* became legendary to the Central Yup'ik.

Reaching farther afield, the Greenland Inuit also use the same word, although pronounced *amikoq*, and to them it means cuttlefish, a tentacled cousin of both octopuses and squids. The Indigenous people of Greenland to the east of Canada and those of Alaska to the west have been in close enough contact over the millennia to share a linguistic history.

That was the story to the west of the Eyaks. To the southeast, along the thumb forming the Alaskan panhandle, words for octopus among the Tlingit, Haida, and Tsimshian peoples all derive from words for bait or halibut fishing, since octopus makes an excellent bait.

Octopus, *tse-le:x-guh*, *amikuq*, *amiguk*, *amikuk*, *amǧux̂*, *aaqanax̂*, and more Alaska Native words from southeast Alaska are all siblings of a sort that linguists refer to as interlingual synonyms. Michael Krauss also mentioned that the word *amiguk* means an old skin covering in the Baffin language, spoken on Baffin Island off Nunavut, Canada. The root sound, *am-*, meaning skin, is widely used in Inuit-Yupik-Unangax̂ languages.

There it is again, skin, at the root of a family of words for octopus. We do not use their hides for shoe leather, fur capes, or footballs, but one essence of the octopus is its skin. This has been important in the evolution of octopuses and their kin from ancestral armored cephalopods, as well as in Alaska Native languages and to biologists using their field keys.

All cephalopods are mollusks: the clams, snails, and related forms. What initially made cephalopods distinct from the rest of the mollusks was their ability to swim, to be neutrally buoyant in the sea, to be mobile. Cephalopods achieved this remarkable agility—remarkable within their evolutionary family, which, after all, contains the slug, our metaphor for decadent idleness—through chambers in their shells. The creatures fill these chambers with gases. When filled with air, the shell and the animal float in the water. Cephalopods were balloonists in the ancient seas. Today, chambered nautiluses still pilot their balloons over the ocean abyss. They hide during the day in deep water and ascend into the oceanographic stratosphere at night to feed in shallow water. The shell gives them the freedom of underwater flight.

Buoyant though it may be, a shell is still a demanding thing to grow and bulky to haul around. Most cephalopods with external chambered shells eventually went the way of human-powered ornithopters and dirigibles. Nautiluses are the last hobbyists of that ilk. But subsequent generations of marine aquanauts (to stretch a metaphor) succeeded by internalizing and reducing the shell, pumping up their respiration. It is in this way the octopuses and squids now swim without the balloon. Here the parallel to flight breaks down—since water is thicker than air, swimming came without airplane-like modifications. However, a few squid do take to aerial flight, expelling water out of their siphons to accelerate to speeds as fast as fourteen knots through the air.

Three features separate squids, octopuses, and their relatives as a group from earlier cephalopods—they lack an external shell, have greater respiration capacity achieved by pumping water over the gills with the mantle, and have complex color-changing arrays in the skin. These features make these animals different from their closest relatives, and set the evolutionary stage for the diversification of squids and octopuses with changeable but distinctive skin patterns.

Many of their predators use vision to hunt. But octopuses are masters of camouflage using their skin, and a hidden octopus is also a safe octopus. An octopus's skin is unusual among animals: the myriad pigmented cells that allow an octopus to change color are controlled by muscles, which in turn respond to the nervous system. An octopus can change color as rapidly as it can move. No slow chameleonlike changes for them. That is how bands of color hastened down the arms of the tiny octopus in a tide pool on the beach. Biologists have dubbed the behavior the Passing Cloud display.

DOWN THE BEACH, Tania appeared from around the bend and waved me on. "Hurry up!"

I was only inches ahead of the rising tide. Although the Sound was calm, each small wave that broke on the steep shore added a cupful more of the ocean to the pool at my knees. I gently lowered the rock back over the diminutive octopus, and stood for a moment, now that she was out of sight beneath the rock. I left her then and wished her well, and started slowly down the beach toward the rest of the day, puzzling over which aspects of her celestial skin display identified her as a very young giant Pacific octopus.

TO DISCOVER THIS, I would finally have to travel to Seattle, Washington, to meet another species. In the Puget Sound Hall at the Seattle Aquarium, a different octopus looked out of the square tank, thirty inches on a side and mounted on a simple stand. The octopus could look across the hall toward the city shoreline, and she also had a view into the briny depths of a truly large tank where rockfish swam. The hall was otherwise empty—just the rockfish, the octopus, and me.

This octopus displayed her differences from the Alaska octopuses, and these were puzzling. Although clearly a similar octopus, she was subtly different. But how, exactly?

All the octopuses I had seen before in Alaska were of a single species, the giant Pacific octopus, *Enteroctopus dofleini*. *Enteroctopus* is the genus of the world's large octopuses. The specific designator honors Franz Doflein, a German taxonomist and crustacean specialist, who in 1904–1905 mounted a research expedition in Japan. Gerhard Wülker, another zoologist interested in mollusks on that expedition, wrote the scientific description of the giant Pacific octopus, naming the species after the expedition leader. The striped sea anemone, a marine snail, some crabs, fish, and a salamander also are named after Doflein.

But this creature in Seattle was the other species reported in South-central Alaska, the Pacific red octopus, *Octopus rubescens*. What set the two species apart?

That she was small was again of little help—every giant octopus must grow through juvenile stages until they exceed the size of the largest red octopus. However, her skin was notably pebbly—a delta of small round islands of paler tissue, slightly raised, each surrounded by darker seas. I had never seen a similar display among the wrinkled folds of the giant Pacific octopus. Those wrinkled folds were entirely absent here. The papillae emerging from her mantle were pointed, yes, but they were spindly—very narrow for their length. They were also distinctly different, now that I had seen them, from the broad papillae that arose like distant snowcapped peaks along the mountain-range mantle folds of giant Pacific octopuses. Under her eyes, where giant Pacific octopuses were entirely smooth, were three small but less-spindly papillae like coarse eyelashes. Another distinct difference between the two species.

I moved around the tank on all sides, seeking a vantage point from which to take the best pictures I could of this little red octopus. I startled her. A lighter band pulsed into place, wrapping from side to side across the football of her mantle. I'd only once seen something similar, but much less distinct, on a giant Pacific octopus.

I was now certain that we were not confusing small giant Pacific octopuses with this other species. The details of the skin were completely different, and recognizably so with the attention we had been paying to the appearance of each octopus we found.

I was developing a more nuanced eye in identifying these body patterns, particularly features that were unchanging within a species but different between them. Other octopus biologists were learning the same lessons in other ways. Octopuses were more often being kept in captivity due to improvements in aquarium science, and were more often photographed alive due to improvements in underwater camera technology. Given the changeable appearances for which octopuses are famous, this increased accessibility to healthy animals allowed us to document their living body patterns in detail.

Now that I had a handle on both large and small octopuses, and better understood how to find their homes, I could begin to unravel another important question: What was limiting or encouraging octopus numbers? As with the question of how large giant octopuses could be, I wondered if Alaska Native stories bore traces of octopuses past. At the same time, I found that other interesting clues about octopus population histories were revealed in nearly forgotten scientific accounts.

Under rare oceanographic circumstances, could devilfish swarm the ocean shores, locusts with eight legs instead of six, but with equally insatiable appetites?

# 5

## Octopuses Overrun

**May 1899, Devon and Cornwall Coasts, England**

L ate in May, a fisherman at Bexhill hauled up the first octopus in a crab pot. It was a big octopus, with a span about as long as the fisherman's arm. It had eaten every crab in the pot, which were only three, but the empty crab shells would bring no money at the market. Still, the notable thing about this octopus was neither its size nor its appetite, but that this was the first that the fisherman had caught in thirty-five years working these waters.

ONE RARE OCTOPUS is curious, perhaps just a vagary. Nonetheless, the odd catch does raise questions. What determines the limits of a species' range? What controls their numbers?

The planktonic hatchlings of many thousands of eggs laid by an octopus mother drift as paralarvae in circulating ocean currents that may obscure both their sources and their fates. The growth or deaths that influence their numbers along the shore could all be happening out at sea, where they were difficult to study, and when octopuses are small and vulnerable.

Temperatures have direct effects on the fates of these paralarvae, influencing metabolism and growth. In the same ways, ocean temperatures also affect their food and their predators. Temperature differences further drive ocean currents in different geographies as the water becomes cooler or warmer. These reverberations of physical properties through biological food webs are known to ecologists as bottom-up forcing, from the base of the food chain upward to grazers and predators.

I CONTINUED TO WALK the beaches of Prince William Sound during my surveys at the lowest tides each year, collecting data on octopus numbers and sizes. Again and again, I found myself curious about aspects of octopus biology that were paralleled by aspects of Alaska Native stories and experiences: The tales of et'stli'yatl, whose size defied credulity, and the impressive size of ocean giants. The destruction of Old Chenega village and loss of its octopus dens in the 1964 earthquake. The loose similarity of octopus evolution and taxonomy to the shifting vocabulary of Alaska Native words for the animals.

There is an Alaska Native story in which numbers of octopuses played a key role. I wondered, did this story echo a deep cultural moment, a time when octopuses seemed about to overflow the sea? And had anything similar been recorded elsewhere? I found versions of the story in several books of legends of the Haida and Tlingit peoples. Like the account from a century earlier in Bexhill, the Alaska Native story began with just one octopus and just one person. And, as we shall see, the eventual arrival of many octopuses on the shore had distant or remote causes.

### Haida Village at Hydaburg Creek, Southeast Alaska

Raven woman knelt above the shore and dug in the earth, gathering roots on an island across the water from the village where she lived. She came here to dig roots more and more. Each time she came, a strange red-haired man appeared to her in the meadow. Together they would dig roots until her basket was full. Sometimes they walked on the beach.

Underwater and intent to reach his destination, an octopus approached the shore. In the shallows, the octopus changed. His back two pairs of arms became legs. His forward two pairs of arms became sturdy graceful shoulders and the strong well-proportioned arms of a man. His hair was red and short, and it stuck up at odd angles on the back and top of his head. He was fine looking, if a bit unusual.

When he appeared fully a man, he pulled himself out of the water. He went to Raven woman, and she greeted her familiar companion warmly.

Today, however, he led her to the shore, to a tiny stone beach protected for a few feet by large rock outcrops on each side. He took her in his arms and they entered the water together. He carried her into the sea.

Raven woman's canoe was found two days later, floating among the islands across from where the village lay. The Chief and his wife mourned for her, as she was their only child, and her people mourned too because she was well liked.

## Winter 1899, Devon and Cornwall Coasts, England

IN THE NORTH ATLANTIC, the winter before the fisherman first found an octopus in one of his crab pots, the Icelandic Low, an atmospheric low-pressure zone, sat over Iceland and southern Greenland. This Icelandic Low was unusually deep, a condition associated with the positive phase of the North Atlantic Oscillation. Relatively low pressure over the North Atlantic resulted in southerly winds and mild winters in England, as the atmospheric high over the Azores pushed northward. The high pushed Mediterranean-fed eddies of ocean currents off Portugal and Spain along with it, until the currents bathed the Isles of Scilly and flooded the English Channel with unseasonably warm waters from the south.

For those on the land, the warmer currents brought a wet and not particularly mild winter. For those in the sea, however, winter stayed farther north than usual, and its chilly grip did not penetrate to the south of England. In surface waters of the English Channel, Mediterranean octopus paralarvae thrived in the mild temperatures.

### Raven's Brown Bear House, Haida Village at Hydaburg Creek, Southeast Alaska

Some time after Raven woman's empty canoe was found, two small devilfish climbed the steps of the Chief's house one morning.

The octopuses reached the top of the steps and entered the Chief's house. They slithered and popped across the floor until they reached where the Chief and his wife were sitting, and then one crawled onto his lap and one onto hers.

"Our daughter lives now with the Devilfish People," the Chief realized. "My grandchildren, is that you?"

Whereupon the little devilfishes put their cold sucker-studded arms around his neck and squirmed about. They twirled their slender arm tips in the Chief's hair, and held to the fingers and hand of his wife.

Then they all went down to the water together, and the Chief and his wife watched as the devilfish entered the water and disappeared under a large rock just in front of the village.

Later, children from the Chief's house were playing on the beach when they noticed two small octopuses. They poked the octopuses with sticks. The Haida people did not allow such games, and told their children to respect everything, and never to abuse living creatures. But these children forgot their lessons, so curious were they about the octopuses, and so caught up in their play. They tried to flip the baby octopuses over to see their suckers and beaks. The octopuses tried to get themselves back in the water and eventually were able to swim away.

Exhausted and a little bruised, their skin dirty and gritty with sand and bits of seaweed from the beach, the little octopuses returned home under the large rock. They told their octopus mother and father what had happened.

As their story unfolded, Raven woman's face grew sober with concern, but their father's face became stony and dark with anger. He turned to their mother.

"How can they be treated so in the land of their own grandfather?"

The octopus people met. They must demand payment for what had been done to their children. They made up their minds to take revenge.

That night, the Raven woman appeared in a vision to Man-that-eats-Leavings. "Warn my father, Raven Chief. There will be a terrible war," she told him. "Keep watch at night."

··········································

### Summer 1899, Devon and Cornwall Coasts, England

BY THE MIDDLE OF JUNE, fishermen from Beer and Babbacombe complained about the octopuses taking crab and lobsters from their pots. As the season advanced, fishermen were pulling up thirty or forty octopuses per day in their gear. Plymouth fisherman abandoned the shell-fishery due to losses to octopuses.

During August, octopuses entered the Dartmouth Harbor in great numbers. They were larger now, having arm spans of four feet or more. Anglers in the harbor caught only octopuses on their hook and lines.

In September, fishermen reported that very small octopuses were prevalent south of Devon. Across the channel, boys in the ports at Cherbourg and Omonville in France caught them in play, but were not able to sell them. At nearby Urville, they were caught by prodding under the rocks at low tide. Fisherman baited their hooks with pieces of octopus arm, but caught only more octopus. So hungry were the octopuses that they struck the bait immediately, and were pulled out one after another almost as fast as the bait could be lowered into the water.

Octopuses remained unusually abundant throughout the winter and well into 1901. By 1902, they were again scarce or absent on the coasts of south England.

··········································

### Haida Village at Hydaburg Creek, Southeast Alaska

The octopuses came ashore. They squirmed on cold arms from the still water after the moon had set, while the people were sleeping. The village had heeded Raven woman's warning. The octopuses reached the village houses and found the doors closed against them and the smoke holes blocked. By this device, some were saved who might have died. But the octopuses entered through every crack into the Raven Houses, and spared only the Eagle Houses. Particularly they surrounded Brown Bear House, where the children lived who had offended them.

Large and small devilfishes crept in the starlit dark and entered into the houses in great number. They crawled into the beds of anyone who was asleep, and lay over their faces, smothering them. Brown Bear House became filled with the slime and dark bodies of the devilfish.

At last, a glimmer of dawn appeared in the sky, and the octopuses journeyed back to the sea. Slipping out through cracks and crevices, they fell onto the ground with plops like the bursting of mud bubbles and slithered in the dawn twilight down to the beach and into the inky green waters.

## 1950, Devon and Cornwall Coasts, England

FIFTY YEARS LATER, the warm winter currents and the octopuses in large numbers returned to the English Channel. In May, hordes of octopuses again decimated the shellfishery, and remained through the summer and fall. This time, the plague lasted three years, through 1952, before notable declines occurred.

### Haida Village at Hydaburg Creek, Southeast Alaska

Some time passed and then the Chief gave a big potlatch. Knowing the error of the children, he sent a message to his daughter, imploring that she and his grandchildren come. The potlatch was on the beach, and the Chief gave gifts of clams and crabs to the octopuses. Man-that-eats-Leavings carved a totem with the image of the octopus, and this was at North Pass for many years.

...............................

## Shimane Coast, Western Japan Sea, 1982

A WARM RING FROM the Kuroshio Current, which formed and spun toward the western Japan Sea in May of 1982, moved northward from the tropical Indo-Pacific. As this child of the Kuroshio Current neared the coast, an abrupt warming occurred in the Japan Sea from May into early June. Argonauts sailed on these currents.

The pelagic octopod *Argonauta argo*, also known as the paper nautilus or greater argonaut, occurs offshore in tropical and warm temperate waters worldwide, where, gulping air under its shell, it rides in the water column near the surface. This animal is a relative of octopuses. The female secretes a papery shell that she holds with her arms in which to brood eggs and regulate her buoyancy. The shell is coiled and similar in shape to a nautilus shell. Although an open ocean species, the animals' fragile empty shells will wash up onto beaches at times.

Adult females are the size of a man's hand. Only female argonauts were found in ancient times. Males were unknown. White wormlike parasites occurred within some females, described in 1829 as the species *Hectocotylus argonautae*. The described parasites were taxonomically odd, however. Observations in 1842 reinterpreted this "parasite" as the males—the whole animal, greatly reduced and modified. Only in 1852 and 1853 did two papers establish that *Hectocotylus* were neither parasitic creatures nor metamorphosed males, but only the detached mating arms of dwarf males.

We now know that the males are very small and do not have a shell at all. Rarely, they are discovered whole, inhabiting the shells of their mates. During mating, the hectocotylus arm may detach from the male. Even detached, this arm is still capable of swimming, clinging, and delivering sperm to the female. Octopuses of all species mate with a hectocotylized arm, its tip modified to aid in sperm delivery to the female. For most species, though, males are not dwarf and the hectocotylus does not detach during mating.

In mid-June of 1982, large numbers of female argonauts were caught by fishermen using fixed set nets on the beaches of the Shimane Prefecture, although they normally were not found this far north. At times, a single haul of the net could bring in several hundred animals. The arrival of the argonauts was accompanied by unusually good catches of tuna, as well as young bonito, pilot fish, pompanos, and squid—all more tropical species associated with waters farther south in the Kuroshio Current. Unusually large numbers of argonaut octopods continued until October in some parts of the country.

IN THESE INSTANCES of the common octopus (*Octopus vulgaris*) along the south coasts of England, and *Argonauta argo* off Japan, unusually warm ocean currents promoted the northward expansion of a species' range into new, usually colder areas, where they were not normally found, or at least, not usually abundant.

Octopuses lay many thousands of eggs. A single giant Pacific octopus may lay over a hundred thousand eggs, a fecundity also found in other species whose young hatch into the plankton. This is necessary, because the decimation of the tiny forms will be huge. Many will be eaten by predators; others will starve; still more will never arrive at suitable habitat. For any nongrowing population of octopuses, only two eggs per clutch survive to mature and reproduce themselves. Numerically, the greatest of the intervening losses occur very early on.

In rare years, however, this balance shifts in favor of paralarval survival and abrupt increases in population. The potential for a great leap in juvenile octopus numbers in shallow water thus is ever-present in the scale of egg laying, although it is nearly always forestalled in this culling of the plankton before the developing juveniles settle to the seafloor.

When I started my octopus work, scholars understood much about temperature changes and their effects on animal populations and movements. But over the decades of my work with octopuses, the effects of climate change on ocean lives would become clearer even to the general public.

# Missing Octopuses Again

# 6

## Global Octopuses

### Cordova, Alaska

A light rain fell from leaden gray skies onto still water at the mouth of the Cordova Harbor, where I looked from my office window. During that first summer of octopus work in 1995, every day's weather offered rain. Climate, on the other hand, takes a span of decades to see.

I rose from my desk and headed to the harbor. A few hours later, I departed aboard *Tempest* on that inaugural diving and intertidal octopus expedition into Prince William Sound. To the west, the clouds broke. Even though it was July, the Chugach Mountains on the horizon sparkled with snow and glacial ice as we sailed into the first and only week of sunshine I saw that summer. I had been on the waters of Prince William Sound already many times during that cold wet spring and summer. In mid-May, I had chartered a seaplane to take me from the western Sound to Cordova so that I could recover from a bout of pneumonia. By the first week of June, I was in an open skiff counting Marbled Murrelets, mergansers, and Harlequin Ducks huddled in Quillian Bay on a cold gray morning as snow swirled around us, melting when it hit the chop. The snow later turned to rain.

In that first year, I worried about how dangerous large octopuses would be, and whether we would find enough animals to study. As I learned to cope with these concerns, I was more able to focus on the central questions of the work: Where were the octopuses? What characterized where they were found? What controlled their abundance? I also located some areas that we would return to again and again, where we could find octopuses with regularity.

At the same time, because a major research initiative at the Science Center involved plankton blooms, I was learning beginner oceanography on the fly. Terrestrial animals are often limited by top predators hunting down the food chain. In the ocean, however, animals, especially invertebrates, are limited from the bottom by algae growth that feeds up from the base of the food chain. Phytoplankton are the photosynthetic algae that drive much of the productivity of the ocean surface waters, feeding the swarms of small zooplankton animals carried in the currents that so influence octopus lives.

Two factors matter to those algae—the availability of sunlight and that of nutrients needed for growth. Sunlight in the ocean occurs most intensely in the surface water, of course. But the organic matter containing nutrients eventually sinks, so that nutrients build up out of the light in dark bottom waters, below a certain depth too dark for photosynthesis. Only where and when coastal waters are well mixed, are both light and nutrients available near the surface.

In the long days of the near-Arctic summer in Alaska, the surface waters warm in the sun while fresh water from land enters the sea at the surface. The warm fresh waters float, and the cold saltier water hugs the bottom. There is no force to mix the two. Early in the season, plankton blooms use up the available nutrients in the lighted surface waters, and by midsummer new growth may stop.

In the near-Arctic fall, the surface water loses heat rapidly, and storms mix the fresh water deeper, breaking up the stratification. The coastal ocean becomes well mixed, bringing nutrients to the surface. This condition persists throughout the dark winter, when surface waters can be as cold and salty as deeper waters.

With the increase of light in early spring, phytoplankton blooms develop on the flush of nutrients that arrived in the previous winter. Without spring winds, the surface runs out of nutrients early; but in a windy and stormy spring, surface waters remain mixed later into the year, and so support a longer period of strong phytoplankton growth.

These blooms support teeming masses of zooplankton grazers and their predators. One such predator can be the tiny, recently hatched octopus paralarvae. These look much like their juvenile counterparts on the bottom, but with stubbier limbs. Too small to effectively jet through the water, they are carried in the currents and cling to the underside of the surface film. They hunt the crab zoea and megalops, early life stages of these crustaceans, that feed around them on the surrounding algae. The more crab they eat, the faster the tiny octopuses grow.

Temperature changes affect ocean currents. Current placement determines the conditions along coastlines, where communities of planktonic juveniles may settle onto the seafloor. Changing conditions may bring plagues of octopuses to the coasts of England, strand offshore argonauts along the coasts of Japan, or eliminate populations entirely. Temperatures, then, act as crucial regulators of how the ocean currents function, affecting not just the flow of water, but the availability of the nutrients on which much of the ocean food chain depends.

Even here, at the base of the food chain, predation matters. When productivity is high, the massive plankton blooms feed the sun's energy up the food chain to larger plankton, to the fishes like herring and Walleye Pollock that feed on them, and further to large predators like salmon, sharks, seals, and cetaceans. When the cold waters rich with sufficient nutrients don't rise from the bottom, however, conditions may change such that plankton is harder to find, and the herring and Walleye

Pollock turn to smaller fish to eat, such as juvenile salmon, and to anything else they can find, including paralarval octopuses.

### June 2004, Prince William Sound aboard *Tempest*

Nine years after that rainy cold summer of 1995, the weather was warm during our annual survey expedition. The pitiless sun pressed onto the flat calm sea. Those aboard *Tempest* jumped into the milky green water to cool down. The scuba dry suits that had kept us warm a decade ago now left us overheated on shallow dives in waters warmed by twenty hours of daylight in the near-Arctic summer. No one could remember such weather ever before.

I thought about the role climate played in the invasions of octopuses that had occurred in the last century in southern England. There and then, at the extreme northern end of the range of that species, unusual warm currents had allowed the population to explode (temporarily) into new territory and in great profusion. The octopuses invading England in times past had been the common octopus, a species adapted to temperate and near-tropical temperatures, moving north in warm years.

Those twentieth-century invasions were advance warning of range shifts. By the first decade of the new millennium, the distribution of the common octopus was more around the British Isles than the Mediterranean. Seventy years from now, in 2090, under likely climate change scenarios, this temperate species will occur around Norway and Sweden in the North Sea, about a thousand miles poleward from the mouth of the Mediterranean where they were likely centered before 1900.

The giant Pacific octopuses of Alaska are cold-water specialists. What effect would this warm weather have here?

At that moment in 2004 aboard *Tempest*, we were taking a short detour, related to the warming temperatures, from our octopus-survey destination. Our skipper Neal Oppen had spotted an iceberg. These originated from nearby Columbia Glacier, located ten miles across the mouth of Valdez Arm and ten miles up Columbia Bay. The icebergs escaped the ancient submerged moraine where the glacial terminus rested until the late 1970s. High tides now lifted large bergs over this obstacle; Neal had to be on the lookout for any that lay in our route. Indeed, even as early as 1989, icebergs escaping this glacier were a problem for ship traffic. This phenomena contributed to the *Exxon Valdez* oil spill, as that tanker turned aside in part to avoid ice in the shipping lanes.

We were headed straight toward one fractured glacier remnant. Above the waterline, the ice was a milky white streaked with black lines and glinting in the sun. A few black-legged kittiwakes, white like the ice with gray and black in the wings and tail, and with natty yellow beaks, rested on the crest of the berg. Below the water, the bulk of the ice appeared shades of turquoise and aquamarine blue before fading out of sight. The chunk of glacier was as wide as *Tempest*. As we glided nearer, the ice loomed higher than the rails of the deck. Cautiously, peering below the waterline to see the extent of the submerged ice, Neal slid us toward the motionless berg. *Tempest* came to rest over one full iceberg-length away from it.

"Ice can roll," Neal said, explaining the respectful distance *Tempest* stood off from the iceberg.

He grabbed the twelve-foot long boat hook, and untied the small aluminum skiff. He pushed off from *Tempest*, and using the boat hook, pulled himself in the skiff toward the ice. Then he stabbed the berg, using the point of the boat hook like a pike pole to break loose chunks of ice about the size of a microwave oven. These he pushed our way. Lifting the netted ice up on deck required the strength of two people leaning over the rail from *Tempest*. The ice in the cooler that held our food was melting fast in the hot sun and we needed more. That evening we cooled our dinner drinks with the gleaming frozen crystals that had formed thousands of years ago from snow falling in the high mountains before beginning their long slow slide down to sea level in a river of ice.

More ice was melting because our part of the world was warming. Warm temperatures restrict ocean mixing by stratifying the water column, while cold temperatures encourage mixing. Recall that well-mixed ocean waters, with nutrients from the bottom mixed up into the sunlight, encourage growth, and that cold-grown plankton population outbreaks can outpace their predators, who often are longer-lived and slower growing than their prey. In such abundance, octopuses grow quickly, and their sometimes-predators feast on other, more abundant, prey. All of these factors contribute to the correlation of cold salty surface waters with populations of octopuses and squids around the world.

The lack of summer mixing between the warm, fresh (and often still) surface waters and the deeper colder saltier water was a problem for us in summer, if we wanted to capture an octopus and keep it alive on the boat. The water at the surface could be so fresh and so warm that it would kill an octopus if we put one in that water, even if we had just picked the octopus up from the shore at low tide. Luckily, the surface layer of fresh water was often thin (this freshwater lens sits atop the ocean brine, particularly after a long rain, or near glacier runoff channels). To circumvent this challenge, we lowered a hose ten to fifteen feet down and drew up water that was colder and saltier than at the surface.

Over the many decades since the emerging octopus swarms in southern England of 1900 and 1950, the planet has warmed—especially in the North Pacific. Giant Pacific octopuses seemed more likely to thrive in the cold, and from our perspective in 2004 these developing conditions did not seem favorable.

### July 2020, the North Pacific

Warm temperatures broke records with dismaying regularity in Prince William Sound, the North Pacific, and around the world. Ocean environments for sea life are warming. Let's begin with March 2015, eleven years from the melting glacial ice of 2004. In that month, over the eastern Gulf of Alaska and over the entire North Pacific, temperatures were the warmest they had been in the prior 136 years of record keeping. Over twelve months, 2015 tied for second-warmest year in Alaska, exceeded only by 2014.

In 2016, the 2014 record was broken; in 2018, the 2016 record fell; then 2019 broke that record. New records were set yet again for 2020, and in 2021 global ocean temperatures were the warmest in history. The new records will soon themselves be exceeded. In my study sites that I had been visiting for twenty-five years on the shores of the Sound, I was no longer finding octopuses.

My colleague Roland Anderson organized an annual Octopus Week every January or February at the Seattle Aquarium, featuring octopus-related events. These included a weeklong volunteer dive survey in which recreational scuba divers reported their encounters with octopuses. The Seattle Aquarium summarized the reports for the public. The first such survey was in 2000. The Seattle Aquarium was kind enough to share these data with me. I compared them to my own octopus counts beginning in 1995, and to publicly available temperature data.

At first glance, the shape of the two curves, octopus abundance and North Pacific temperatures, looked like mirror images. They weren't, quite—there was a lag. When octopus counts had been high, temperatures were not low that same year but had been low a year or two before. The lag corresponded to the approximate age of the octopuses counted in the survey. With this lag, I drew a graph with temperature on one axis and the octopus count on the other. I saw a sharp downward sloping line—as the oceans warmed, octopus counts fell. This was not just in Alaska, but also for the Puget Sound octopuses counted by the volunteer divers for the Seattle Aquarium.

Warming waters resulted in fewer octopuses. Based on the necessary lag, the temperature effects on the octopus population likely occurred when each cohort was paralarvae in the plankton. Warmer waters made it less likely for paralarval octopuses to arrive and settle onto to the beaches where I counted them or into waters up to sixty or seventy feet deep where divers found them.

For the giant Pacific octopus, as well as other octopus species, there is a thermal neutral zone, within which temperature changes do not much affect octopus well-being. In the thermal neutral zone, the metabolic costs of dealing with temperature change are not high. When temperatures exceed this zone, the energy costs of dealing with even small temperature changes can be very high. One study found that octopus survival fell by 15 percent with warming just a degree beyond the neutral zone; and with larger temperature changes survival declined by 70 percent. Ocean warming is already leading to geographic range shifts poleward for some octopuses, and continued warming will lead to dramatic population declines in the near future.

In June 2020, I revisited our study sites, hoping to collect some octopuses for new research in the aquariums. I chartered a small float-plane. It was the first summer of the COVID-19 pandemic and social distancing prevented me from hiring a boat or bringing a research team of university students. No matter. The best octopus seekers in Alaska were ready to go with me—my children had grown up combing the beaches for octopuses and were already part of my social bubble.

We headed out into blue skies over the Chugach Mountains from Anchorage to Prince William Sound, to visit the beaches we had previously reached from the research vessel *Tempest*. We would get only one crew on one beach per tide. Our sampling would necessarily be less than in previous years. I focused on bringing a few healthy octopuses back. However, I was to be disappointed.

On two different flights, we visited our two best beaches. In the early years, we had never missed finding signs of octopuses at these sites. At times we found so many occupied dens that the returning tide rushed our surveys of middens and habitats, which we had to complete before the dens and our entire field site were submerged. In the latter years of our work, we were finding fewer octopuses, but their favorite dens were often still occupied, particularly those in the lower intertidal reaches.

In 2020, however, I was shocked to see almost no sign of octopuses at either site. On the first day, we found only one den that had been recently occupied, but perhaps no more recently as sometime earlier in the year, late in the winter. There was no one home the day we checked.

At the second site, we again found only a single den where an octopus had recently been. It was perhaps as recent as earlier in the spring, perhaps a month prior to our visit or less. Again, there was nobody home.

Early in 2021, with the help of another colleague, I was able to include a few more record-setting years of climate change, and evaluate a different set of octopus counts from divers in Washington State who were participants in the Volunteer Fish Survey Project for the Reef Environmental Education Foundation (REEF). The REEF counts paralleled the results of Octopus Week from the Seattle Aquarium, and my own data in Alaska—in each data set, fewer octopuses appeared follow-ing warmer years.

I began my quarter-century of fieldwork in Prince William Sound with concerns about the spilled oil from the *Exxon Valdez* and its effects on octopuses and their habitats. I ended with similar worries as the first signs emerged of the current and anticipated effects on octopuses from climate change and the warming oceans. Ever present over this time-span, however, is the challenge to understand the effects of harvest-ing octopuses.

# 7

## Octopuses Seized

Andavadoaka in Madagascar is a study in contradictions—a coastal village that hasn't seen rain in three years, where cell phones are common but basic plumbing is absent, and where reef gleaning, a traditional livelihood for women, is now supplying a global trade in harvested octopuses.

Glimmers of dawn reach through five feet of slightly cloudy blue water. The seafloor below is scattered with bone-colored sand and patches of coral skeletons overgrown with short brown algae. Tide is falling. A curtain of monofilament net stretches out of view in either direction. In the net, otherwise without catch, hangs one dead Red Squirrelfish, protein at most for a single serving. Farther, the reef flat gives way to sand and eelgrass beds before the next reef section. As the tide falls and day brightens on that reef flat, an octopus awakens. Each of her arms alone has more meat than that entire squirrelfish. She is sheltered in a hole in the reef rock. There is no pile of food remains outside her den, only a small spray of unremarkable debris where she broke off a piece or two of the reef rock to make the inside of her space more comfortable.

Above the octopus, through no more than several inches of low tide, a woman approaches on bare feet over the sharp reef. She lives in Andavadoaka and is Vezo. The reef flat is within the traditional gleaning waters of her people, those who make their living from the sea. Vezo are not of a particular ethnic heritage, but adopt their identity with their way of life and their community's traditional fishing grounds along Madagascar's west coast. Also among the Vezo are those with *renetane* lineages, first occupants of their motherland.

The woman's faded fuchsia skirt is twisted about her knees to keep above the water; her face is caked with *tabaky*, a pale paste of ground wood, bark, and water—already dried, cracked, and flaking off—to protect her from the sun. She carries with her a length of rusty rebar and a wooden stick, its end sharpened to a point and hardened over hot coals. She stops at the hole, and after checking, deftly inserts both the wooden pike and the rebar. Within moments, the octopus erupts from the small opening, seeking escape like dark storm water from an overflowing drain. The reef gleaner picks up the octopus, dispatching the writhing captive in an instant by turning her suckers out and pithing her through the mouth and brain. The octopus goes limp.

It is an *Octopus cyanea*, the day octopus, a medium-sized species found from the coasts of East Africa across the Indo-Pacific to Hawaii. Artisanal fishers throughout their distribution catch them for subsistence or sell them in commercial harvests. The meat from this octopus will be sold and shipped to the northern hemisphere to feed the eager markets of Europe. It may provide the only cash income for the Vezo woman's household. Traditionally, her community did not eat octopuses—it was only after the commercial market for seafood arrived at Andavadoaka shores that octopus harvesting became important. Today, her family will have to find their own meal elsewhere on the reef.

Andavadoaka lies just over one hundred miles north of the nearest city, Toliara. The first twenty-seven miles are on paved road, less than an hour by car. The remaining distance requires six hours to traverse under favorable conditions, as corrugated dirt paths bounce and ricochet the car through the spiny forests. Traffic in the village itself is on foot or with carts pulled by zebu, domesticated descendants of the now extinct wild Indian aurochs.

Madagascar is one of the poorest countries in the world. Seventy percent of the Andavadoaka population make their living on the ocean—people here rely heavily on the sea for their food, their livelihoods, and their identity. The village, lying on the southwest coast of Madagascar, gets little rain. The spiny forest surrounding the village is dry. There is no soil but sand. The Vezo practice no agriculture, although neighboring inland people grow some crops. Dry goods such as flour, rice, pasta, and lentils keep well and are brought in along the road. Fresh vegetables are scarce. Goats graze the dry vegetation and provide some dairy and meat, but most fresh food comes from the sea.

The reefs of Andavadoaka have declined in recent years, and now have only 20 to 30 percent coral cover typically, and low fish abundance and biomass. Where coral cover is low, algae can invade. Grazing fish are important to control algae and promote coral reef health. Their abundance on reefs is threatened when their predators are absent. The larger coastal species of fish that people prefer to catch, such as coral trout, often are predators on smaller grazers. When the larger predators are gone, people harvest the smaller grazing fish. On the fringing reefs along the shore, the impacts of the Vezo fishers are apparent.

One day, I was paddling in a pirogue with Bris, an Andavadoaka local. Without warning or preamble, Bris tipped over the side into the water, grasping at his wooden octopus pike on the edge of the boat as he went under. Bris had agreed to take me with him as he went collecting octopuses in deeper water by breath-hold diving. We were headed back to the village after a day without finding octopuses. I brought the pirogue to a halt, and peered down into the clear blue water. Bris hung there several feet down but off the bottom, holding the sharpened stick like a spear. He was at home underwater, turning each way peering around him. When he came up for air, he explained why he went overboard.

"I thought I saw a big trout! I wanted to catch it. They used to be common here and we would catch them, but now we never see them."

The coral trout is a type of grouper, and they can grow to over a meter long and up to fifty pounds. Their meat is firm and clean white, and would feed many. Once underwater, Bris had been unable to catch a second glimpse of the fish he thought he had seen from the pirogue.

Over recent decades, many large fish became scarce. Purse-seiners from the European Union and Asian long-liners have increasingly harvested tuna and billfishes offshore from these waters. Domestic catches of shrimp and other seafood have also grown over this time. International seafood collection and export companies brought accessible commercial markets for large pelagic and reef fish species, as well as for the octopuses. Climate change and population increases further threaten the nearshore marine systems that sustain the Vezo.

When the preferred large fish leave a reef or were depleted, fishers must turn to other game, typically smaller fish and invertebrates. These can include migratory schooling fish and small grazing fish that clean algal growth off the reef and maintain healthy coral cover. As the international export market for octopuses and other fish has grown, the catch size and numbers have declined.

Bris and I dragged the pirogue onto the beach at Andavadoaka. Young men and children sitting or playing along the village edge ran down to meet us. Together we carried the pirogue higher up above the tide line to rest in a row of similar boats. As I walked along the beach in the hot midday sun, I passed wooden racks of sardines. The fish were about five to eight inches long, caught the night before and now drying in the sun. Schools of sardines near the village, too, are seen less frequently now than in the past.

I paused as I was about to skirt a patch of sand that was dappled with tiny silver fish, as slender as a pencil and less than two inches long. What were these?

The little fish seemed too small to eat, and further, were lying in the sand, apparently too small even to put out on racks. They appeared at times in schools near shore. The mass distribution in Africa of fine-mesh bed nets as protection from malaria-carrying mosquitos was supposed to reduce disease. In a practical fashion, however, the nets were immediately repurposed to catch schools of tiny shrimp or fish, such as I now saw drying on the ground. Once dried, the fish were shaken free of sand and used in soups or bartered with neighboring inshore people who did not have access to the sea.

As the Vezo fishing community became aware of these declines in their harvests, they took action. They were aided and encouraged by Blue Ventures, a UK organization dedicated to sustainably improving food security in ways that make local economic sense. In 2004, traditional reef gleaners, frustrated by finding fewer and smaller octopuses, temporarily closed to all harvest parts of their octopus fishing grounds—they had voluntarily protected locally managed marine areas. They named their management areas Velondriake, "to live with the sea."

Octopuses are susceptible to overharvest. With only a short period of harvest in dive fisheries, for example, catch rates begin to decline. This is a local decline, when and where the rates of harvest of larger octopuses are faster than rates of octopuses settling out of the plankton and surviving to harvestable size.

The good news is that *Octopus cyanea* can recover quickly. This is a fast growing species that can double in size in just over a month. Seven months later on returning to harvest the temporarily closed areas of Velondriake, Vezo discovered abundant and larger octopuses. The sustainable harvest from a managed protected area was higher than it had been without management. The renewed harvest more than replenished the income deferred by the closure.

Neighboring communities noticed the Velondriake success story, and a rotating schedule of temporary closures spread up and down the coast where octopuses were harvested and sold. Vezo must fish every day to feed their family. Neither a short-term closure of the whole fishery nor closures of fixed and permanent protected areas were feasible in such circumstances. However, closures cover only a fifth of a villages' fishing area, and harvest can proceed elsewhere in the remainder. When the closed area reopens, octopus harvest surges upward and then gradually declines again until the next closure.

In ninety countries around the world, artisanal octopus fisheries provide protein and cash to local harvesters who depend on the sea. Local people harvest octopuses for food and commercial trade from the equatorial waters of Madagascar and around Africa, across the Indian Ocean to Thailand and as far south as Tasmania in Australia, through South Pacific islands and throughout Asia, north to Japan and Alaska, and in Atlantic waters of the Caribbean seas, off Venezuela in South America, and in the Mediterranean and other European shores.

Predators can have beneficial effects on their biological communities overall, and octopuses are no exception. Wherever octopuses thrive, as they do in the locally managed marine areas of Velondriake, the result is healthier fisheries and healthier reef communities.

In the world's most bio-invaded marine waters, the Mediterranean Sea, the common octopus may have a role in controlling invasive lionfishes that are predators on many local indigenous fish. This is an example of the possible role of octopuses as predators in marine community health. Lionfishes are generalist predators and are now widely invasive both in the Caribbean and Mediterranean Seas. They came to the Mediterranean Sea likely from the Red Sea via the Suez Canal. And they have been increasing rapidly since their arrival. Due to their large and venomous fin spines, lionfishes have few known predators outside of their native ranges in the Indo-Pacific. Despite their spines, in 2021 an octopus was observed enveloping a lionfish in its arms and web and subduing it. Could an abundance of wild octopuses, appropriately protected from harvest, reduce an abundance of invasive lionfishes?

For many harvested octopus species, there is a general lack of information of the kind most needed in fisheries management. Methods to age octopuses are available for only a few species. The factors determining survival are challenging to study, involving oceanographic and ecological forces that may themselves be distant from where the surviving octopuses are harvested. There are many problems still in getting accurate reports of how many octopuses are harvested, of which species, and from where in the world.

To relieve fisheries pressure on octopuses, some are trying octopus farming. Aquaculturists now raise hundreds of marine species in captive facilities, from tuna to clams. Such farmed products can constitute half of the seafood market in some countries. Farming octopuses appears promising, but it is also challenging in several ways. The promise is that octopuses can grow rapidly and can efficiently turn their food into octopus flesh. The challenge is that octopuses do not appear well suited to captive mass production.

Octopuses are carnivorous, like many farmed marine species. They need animal protein. The catch from one-third of global fisheries is turned into food for other animals, and half of that is fed to farmed species, often in a pelletized form. To date it has proved difficult to develop a manufactured food on which octopuses will grow as rapidly as on live prey. For a variety of reasons also, most marine animals need marine-sourced foods, not terrestrial. Octopus chow likely cannot be made from low-value by-products of the meat agriculture industry.

Octopuses in aquaculture enjoy crabs, clams, shrimp, or other foods harvested from the sea. It would be most efficient to use these foods directly for human consumption, given that they have been harvested, rather than for animal consumption. It makes little ecological sense to harvest wild seafood to feed to carnivorous farmed species, including octopuses, compared to farming planktivorous or herbivorous species that thrive on plants and algae. Despite this, the aquaculture of such flesh-eating species is growing rapidly.

Octopus farming also would not be like caring for an octopus in a public aquarium—where octopuses are given space, diverse foods, and lots of engagement. In aquaculture, many animals must grow rapidly, typically in crowded quarters. Most wild octopuses are typically solitary, and do not thrive this way. They may be stressed to encounter one another. Yet they are very curious animals that thrive when engaged in complex habitats that provide stimulation and opportunities to explore and forage. Octopuses need these activities to be healthy and grow rapidly. Further, octopuses may attack and kill each other when pushed into close quarters, especially if stressed. However, it is worth noting that, in at least a few species, cultured hatchlings adapt to crowded quarters, as long as individuals are all of similar size. Critics have assumed that farming most octopus species might require separating them into small quarters. It will be interesting to see if this remains the case for fully aquacultured lineages. If so, the necessary stimulation and opportunities to explore could not be provided in separate small quarters at commercial aquaculture scales.

Increasingly, scientists and the public are aware that the welfare needs of terrestrial farmed animals are not being met, which is part of the motivation for many people to turn to a vegetarian diet and plant-based meat substitutes. Such welfare concerns will be worse, not better, for the carnivorous octopuses. They are mostly solitary, and yet recognizably intelligent and curious animals. The commercial markets that make octopus farming potentially profitable are predominantly in food-secure nations. The lack of farmed octopuses will not compromise food security. There is no need to develop another farming industry on a model of production that compromises on considerations for animal and environmental welfare. Moreover, octopuses as a farmed seafood species present challenging conflicts with ecologically sensible and sustainable farming practices and animal welfare standards.

Which makes the management of wild octopus habitat and fisheries all the more important, and most particularly so for small artisanal fisheries that sustain coastal communities. Harvest closures that protect octopuses and their habitats do not mean the fishery will be less valuable. As the Velondriake project demonstrates, temporary closures by the Vezo with their locally managed marine areas resulted in both more octopuses and in overall higher value from their octopus harvests.

# 2

## Want

# *Tracking Octopuses*

# 8

## Octopus Scraps

**Underwater in Prince William Sound, Alaska**

he first sign was a single oval, orange crab claw with black finger tips. It was from an Oregon rock crab (*Glebocarcinus oregonensis*) and just half the size of my pinkie fingernail. I followed a scant trail of these remains—small scallop shells and broken bits of crabs. Rising gradually up a slope, at the base of a boulder, I found the entrance to a den and more of the meal remains an octopus leaves after eating. The octopus was on top of the boulder with one arm wrapped neatly around a kelp stipe. Alert and watching, she waited to see if I would look up and notice her. When I did, she pushed into the emerald green water, jetting away.

I followed, swimming fast. I was intent not to lose sight of her. I wanted to bring her back to the boat to measure weight and size, and verify sex, before returning her to her den. She broke to the right. My gaze followed, but she cast a sepia cloud that trailed one coiled tendril and made visible some eddy of the water. Jetting beyond it, she turned sharp left. I was expecting something like this—I pursued through the ink cloud and saw her dart into a patch of kelp. She froze. She turned translucent burnt sienna, the color of the brown kelps. Her skin roughened and became the texture of the kelp. My eye could only find her among the fronds if I focused on her suckers, on her eyes.

Octopuses are perfectly camouflaged and well hidden in kelp or under rocks. But giant Pacific octopuses in Alaska also are messy and leave remnants of their meals behind. I found this octopus by tracking her discards. The most distinctive scraps that octopuses leave are the hard shells of the prey they catch. In Prince William Sound, their favorite hard-shelled food seemed to be Oregon rock crabs. These little crabs have a maximum recorded size of 5.3 cm—the width of a credit card—across the widest part of their carapace, but most of the ones the octopuses caught were only half that size.

Most midden piles left by octopuses contained remains of Oregon rock crabs, more often than those of any other prey species. The octopuses were addressing their own version of the "omnivore's dilemma": What's for dinner? Doing so raised a related question: *Why* is that what's for dinner? Why was the world's largest octopus species so often choosing these small crabs?

Michael Pollan posed the "omnivore's dilemma" in his 2006 eponymously titled book. According to Pollan, people, as omnivores, the most unselective of eaters, face a variety of food choices, resulting in the dilemma of how to find and recognize what to eat, and avoid potentially harmful foods. Indeed, a 2012 study found that hunter-gather humans in Alaska were not picky eaters—they were super-generalists in diet. The Aleutian Islands are the traditional lands of the Unangax̂ people—seasiders or people of the passes. Their language, Unangam

Tunuu, is a distinct branch of the Inuit-Yupik-Unangax̂ language family. The remains of their food preparation and meals, deposited in middens by the Unangax̂ over a five-thousand-year period, revealed that they ate fifty different species from the intertidal alone, more than a quarter of the available species in that habitat, and more types of prey than any other intertidal animal in that study.

Humans are omnivores, eating both plant and animal foods. But octopuses are not. Octopuses are carnivores, eating only animal prey. But are octopuses picky eaters? Superficially, they seemed almost unselective, eating whatever animals they found. In the intertidal waters of Prince William Sound, our midden samples collected over a period of just *twelve* years, which is just a fraction of the timespan sampled for the human super-generalists, revealed that octopuses eat at least fifty-two intertidal species. That was two species more than the human hunter-gatherers of the Aleutian chain. The most common species in Prince William Sound octopus middens were these little Oregon rock crabs.

This was not the case everywhere, however. Giant Pacific octopuses in other regions captured larger prey. In the Salish Sea of Washington and British Columbia, 45 percent to over 80 percent of the diet was the decently large red rock crab, *Cancer productus*. In Aleutian waters, octopuses feasted on two large bivalves, the horse mussel *Modiolus*, and the Alaska falsejingle *Pododesmus macrochisma* (a relative of scallops), which together comprised well over half their diet. Both mussels and Alaska falsejingles can anchor themselves to the seabed with byssal threads, strong fibers made of protein and secreted by a special gland in the bivalve's foot. It takes work for an octopus to unmoor these prey for a meal.

In Prince William Sound, five prey comprised most of the middens. The most common species in the diet, the Oregon rock crabs, made up a quarter of their prey. The bigger red rock crab, almost the sole diet in the Salish Sea, represented only 20 percent of the diet here. Third, the large, golden and fleet-of-foot helmet crab, *Telmessus cheiragonus*, comprised another 17 percent of midden items, but barely appeared in the diet outside of Southcentral Alaska. Fourth, the graceful kelp crab, *Pugettia gracilis*, that bears on its rostrum an elegant and carefully selected blade of kelp, made up 12 percent of the Prince William Sound diet. This species also showed up in the samples from the Aleutians. Fifth and finally, remains of the black-clawed crab, *Lophopanopeus bellus*, amounted to 7 percent of the midden items in Prince William Sound, but these hardly showed up elsewhere.

In combination, all of these five crab species represented 81 percent of the diet of giant Pacific octopuses in Prince William Sound, the same proportion accounted for by only a single species of prey in Saanich Inlet, British Columbia. The first clue to understand why the Prince William Sound diets were full of small and diverse prey types came as we counted the live prey on the same beaches where we surveyed octopuses and their midden piles. The most common crabs found on the beaches were either the black-clawed crab or the graceful kelp crab. But the most common remains in the middens were the Oregon rock crab or the helmet crab. Octopuses were selective.

The crabs they killed were bigger than the live crabs on the beaches. Octopuses have predilections. Live graceful kelp crabs and black-clawed crabs were small, with carapaces that averaged less than the width of my thumb (1.5 cm, or just over half an inch). Those caught by octopuses, however, were typically a third larger or more. Red rock crabs and helmet crabs grow much larger than the other three crab species in the octopus diet, with carapace widths up to eight inches for the red rock crab, and half that for the helmet crab. Carapaces of red rock crabs caught by octopuses exceeded the size of live crabs on the beach by the width of a golf ball (about 4.5 centimeters); helmet crab remains were larger by just over half that difference. Overall, octopuses liked the larger crabs for prey, both big species, and big individuals within every species. They left only the smaller prey behind.

With this answer, another puzzle emerged. Octopuses caught Oregon rock crabs and black-clawed crabs of similar size. Yet the octopuses more often chose Oregon rock crabs, and more often left alive the black-clawed crabs. Crab abundance varied somewhat by the habitat surveyed. Yet on each of our survey beaches, octopus middens better represented the Oregon rock crab than black-clawed crabs. Why was the Oregon rock crab preferred?

Crabs, it turns out, have parasites. In particular, the black-clawed crabs are susceptible to parasitic castration. We first realized this when an unusual number of the black-clawed crabs appeared to be egg-laden. The arrival of spring found female crabs carrying eggs under their abdomen. Aside from the eggs, we learned the sex of a crab by its curves. Underneath the crabs, the female's abdominal sections are wide with convex curves where she cups her eggs during incubation. In males, the curves sweep concavely narrow toward the tip.

What at first we mistook for eggs were not the reproductive efforts of the crab. These were instead the external reproductive sacs (externa) of parasites that infected males and females alike. This parasite, which is a relative of barnacles, grows an internal rootlike system of ramlets (the interna) that extend throughout the body of the infected crab. The parasite externa and interna connect by a stalk. There is no shell, eyes, nor legs, but instead only the parasitic-twisting filaments growing like a fibrous cancer through the tissues of the crab. The parasite absorbs these tissues to provide the energy to make her eggs.

Many such species of these parasites occur worldwide in crabs that inhabit deep and shallow seas, as well as in semiterrestrial and freshwater crabs. The species of parasite infecting our black-clawed crabs in Prince William Sound had been identified based on similarities to an Atlantic species. More recent taxonomic assessment, however, considered this mistaken: the particular species infecting crabs in these areas is not yet known.

Infected crabs stop growing and do not reach reproductive maturity. These crabs seldom grow a carapace broader than the width of a finger, while the black-clawed crabs captured by the octopuses were half again as large. Octopuses avoided eating the infected crabs, whether by their size preference or some other means, which was smart, because the energy content of stunted crabs averaged just over half that of equal-sized healthy black-clawed crabs. By avoiding the infected crabs, the octopuses also avoided low-quality food.

The hairy crab, *Hapalogaster mertensii*, is another species of interest. Bristles cover the carapace, claws, and legs. Octopuses avoided hairy crabs, which were unusual to find live on the beaches, and were exceedingly rare in octopus middens. The meat of the hairy crab is as energy-rich as that of any other crab. They are also soft and slow; and so they should have been easy prey. Avoidance hinted that octopuses were not finding these crabs—perhaps because the octopus were not foraging through the soft sediments under the rocks where these crabs like best to live, although that's uncertain.

Crabs can avoid notice by octopuses, sometimes in elaborate ways. More than once, I knelt on a beach amid swarms of black flies to note the midden pile of food remains scattered across an excavated berm by a den opening. The berms, recently pushed out from below, were bare of overgrowth, but the rock of the dens hung with limp blades of brown seaweed and green or brown tangles of filamentous algae. These I pushed aside, clearing the mouth of the dens. Was an octopus in there?

As I peered in, focused on the depths of the burrows, intent on getting a glimpse of a sucker or eye, a tuft of the algae to one side would move fractionally, getting itself out of my way. The movements were yet another crab species, more adept than most at escaping notice. The graceful decorator crab, *Oregonia gracilis*, festoons its entire body—rostrum, carapace, chelae and legs—with an immoderate collection of items from its habitat. The items are primarily algae but also sponges, bryozoans, tunicates, anemones, and whatever ostracods, worms, skeleton shrimp, or brittle stars clung to their other decorations. The decorations overwhelm the crabs, and until I learned better how to see them, the crabs went unnoticed unless they moved. I found them many times at the entrances to occupied octopus dens. The octopuses did eat them, but rarely.

The giant Pacific octopuses of Prince William Sound capture all five of these crab species—we found at least one of these in almost every midden. Despite the large numbers of crab remains, however, these middens *also* had other species nearly two-thirds of the time. After the five crab species, the next most common prey to occur in these mixed middens were six different species of clams or bivalves. The most frequent of these was the Alaska falsejingle. This thin-shelled rock oyster is common in Aleutian waters where it is a major part of octopus diets and is conspicuously abundant in the environment. A cockle (*Clinocardium nuttallii*), the butter clam (*Saxidomus gigantea*), and the Pacific littleneck clam (*Leukoma staminea*) were next most common. The thick-shelled cockles and butter clams are large like the jingles. The littlenecks are probably abundant in the habitats. The stained Macoma clam (*Macoma inquinata*) and the Pacific blue mussel (*Mytilus trossulus*) rounded out the list of bivalves in the middens ahead of the rare and avoided hairy crab.

The jingles and the mussels anchor themselves to the substrate, the latter quite high up in the intertidal and the former somewhat sparsely. The other four clams typically bury in the sediment. Octopuses may encounter them most often when digging their dens. The further remaining dozens of species of prey captured by octopuses were a mixture of other crab species, clams, the occasional large snail, chitons, and even urchins.

The fresh remains of prey represent a single meal or foraging bout, or at most a few. Octopuses do not always discard meal remains immediately outside a den entrance. Sometimes octopuses eat away from the den, and we found small piles of remains in such lunch spots.

Minor portions of octopus diet would not leave evidence in the middens. Octopuses catch shrimp, whose shells are very fragile and light. I once found an octopus eating a Crescent Gunnel. The octopus had torn away the flesh neatly around the spine about a third of the way along the length of this eel-like fish, the same way we eat corn on the cob and leave behind the inedible core. Most octopuses shred the soft tissue of their prey to bring it into the mouth and swallow. Stomach contents are finely broken up.

Octopuses dine on more than the largest available crabs. They are in fact eclectic in their tastes, as shown by the curious inclusions of occasional items in the diet, such as the Crescent Gunnel. Beyond fish, their dietary choices were sometimes astonishing.

One day in 2012, pedestrians along the Ogden Point Breakwater in Puget Sound noticed a Glaucous-winged Gull on the water in the shallows. The gull was acting strangely. Its head was underwater, but it was beating its wings on the surface. Around its neck was the red arm of an octopus. The gull flapped, but to no avail. The rest of the octopus had a good hold on the rocks of the breakwater. In a few minutes, the octopus had pulled the gull below and enveloped the bird in its web, leaving just the extended wing tips visible. No one saw the start of the encounter. Possibly the gull had been pecking at the octopus in the shallows, mistaking its arms for a worm. But maybe not.

Brazilian researchers in the São Pedro and São Paulo Archipelago in the West Atlantic watched a tropical octopus reach up from a tide pool to grab a Brown Noddy that had alighted on the rim. The octopus drowned the struggling bird and proceeded to feed on it over the next seven hours. In these habitats, octopuses remain hidden in their tide pools, obscured by the water surface or underwater growth. Sally lightfoot crabs (*Grapsus grapsus*) forage on the pool rims, true to their name, and *quick*. They skitter around the rocks, always keeping on the far side from observers or threats. The same researchers watched as a sally lightfoot foraged along the pool edge and came within the reach of a hidden octopus. In a fast grab, the octopus flung one arm from the water and caught the crab above. Another arm followed and ensnared the crab, and together the arms pulled it underwater.

Octopuses are large predators in their realms, and can be fast. Their powerful arms have a wide spread with suckers that grip anywhere along arms' length. The suckers adhere on contact. Curiosity is built into the octopuses' anatomy. Like a human baby that puts whatever it finds into its mouth, an octopus investigates what it touches by passing it sucker-to-sucker toward the mouth. After all, it might be edible. The middens in Prince William Sound revealed curiosity of this kind. Octopuses were hunting large crabs, but at the same time, sampling what the habitat had to offer. If it seemed interesting, or edible, they tried it. Often that was a clam. Sometimes it was a spikey sea urchin, a slippery fish, or even feathered prey caught unawares.

We had an answer to our original question, "What food do giant Pacific octopuses choose?" Mostly here in Southcentral Alaska, they take the five crab species mentioned. They like the larger species, and within species, the larger individuals. Elsewhere, octopuses select narrow diets of one or a few big species. But what accounts for this difference? That is, why are Southcentral Alaska diets more diverse, with more small prey? While there are not many data on the abundance of the live prey in other regions, our live prey surveys in Prince William Sound do provide such data and show that the largest crab species were rare in our study areas. At the northern edge of the range of giant Pacific octopuses, in near-Arctic waters forever altered by the impacts of the massive *Exxon Valdez* oil spill, big meals are hard to find. Without big meals, octopuses catch the little stuff, the only food available. Even here, though, they continue to be smart and selective, taking larger prey and avoiding low-quality food.

This explains the "Why?" of the regional difference in octopus diets. In habitats with fewer large prey, octopus food choices have to be more catholic, just for the hungry predators to get enough to eat. These answers beget new questions: What makes the hairy crabs unavailable or unappealing? How do prey species avoid or escape octopuses? And for that matter, how do octopuses recognize their prey?

To answer such questions, I needed to understand three things: the ways octopuses defeat crabs and bivalves, which are the most heavily armored animals in the sea; how the crabs themselves attempt to thwart the marauding octopus; and what is the sensory world of the octopuses that allows them to detect and pursue a crab intent on escaping their suckered clutches. As it so happened, the *nearly* impenetrable armor of a defeated butter clam, strong enough, thick enough, and almost hard enough to resist attack, would show the way.

# 9

.............

# Octopus Tools

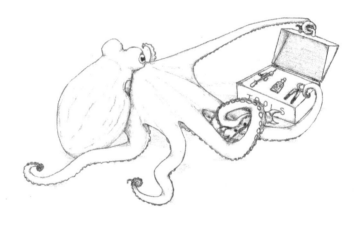

**Shores of Port Graham, at the Mouth of Cook Inlet, Alaska,**
**within the Traditional Lands of the Sugpiaq**

How did the octopus defeat the clam? I picked up the large shell. No other clam in this area grew as big as the butter clam (*Saxidomus gigantea*). The shell was nearly four inches in width, thick, and heavy in the hand; the strong armor of a well-protected animal. Yet an octopus had gotten inside—I'd picked it up from an octopus midden outside a den on a rocky beach.

To get to the village of Port Graham, Tania and I took a cargo jet from Cordova to Anchorage, arriving at a large and modern airport. To get from Anchorage to Homer, we boarded an eighteen-seat twin engine Otter, a large prop plane; and for the final leg from Homer to Port Graham, a four-seat floatplane. As we moved into successively smaller planes, I felt I was drawing closer to the Alaska I had imagined when we first moved to Cordova—an Alaska of legend and dreamtime.

The night before I found the clam, Tania and I had sat in a darkening room, and listened to Simeon Kvasnikoff tell us the history of his family, and a little of the Sugpiaq. We sat at Simeon's dinner table, looking out the picture window across the water. Shadows lengthened across the bay. They crept down the steep mountainside cloaked in brilliant green hemlock and spruce, and with a touch turned each tree deep forest black. On the near shore, the water glowed red in the long rays of the afternoon March sun. Over the window was a faux-wood engraving of Christ and the apostles at the Last Supper, the exact depiction that had hung in my own dining room when I was a child, over which my brothers and I used to tuck palm fronds on Good Friday.

Slightly portly, Simeon was sixty-one, although he looked younger, with a round face, a nearly perpetual grin, and an amused crinkle around the eyes. A cigarette always dangled from his mouth, although he never sucked at it. It slowly burned down and bobbed as he talked, and although I never saw him remove the butt or light a fresh cigarette, whenever one had burned to the filter, another would mysteriously take its place. His black mustache was shot with gray and stained yellow from cigarette smoke. Beneath the cap that he was seldom without, his black hair was also graying, with one lock dyed orange.

Simeon looked at us, sitting opposite him at his kitchen table. "I took the kids there," he said as he gestured out the window toward the mountainside, "to see how the ancestors got the bears out of their holes in the winter. No one believed anymore that we used to do that, so I took them up there to show them."

For furs? Meat? A demonstration of bravery and skill? I didn't ask Simeon why the ancestors did that. The story was the point.

"You see the bears when they bring grass to their holes, going back and forth. That's how you know where the holes are. When I first look in, the bear was facing me. After I move out the grass though, he turned 'round with his head in the back. I went in there with a pistol and shot him in the head. The kids pull him out. The kids, they brought their rifles. I ask them, 'What are you going to do with those in there?'" In the fading twilight, I could still see him smile. "All I had was my pistol."

He talked about his mother. His voice was soft and slow, punctuated now and then with a chuckle. "My mom, she was strict, but she raise me up right." By the time Simeon was a young man, Western ways were encroaching on the villages, driving out some of the old ways. Simeon's mother, however, "wouldn't eat no canned food," and so Simeon became her provider after his father died, for many years supplying her with the wild foods she liked. He fished salmon in spring and summer, caught halibut when he could, hunted seal and ducks in the fall, and crouched silent and still for long hours in the woods waiting in ambush for deer and other game. He picked bidarkis, collected octopuses, and gathered clams from the beaches when the tides were low. Simeon was one of the most active gatherers of wild foods.

Evening twilight had faded and the stars came out one by one into the clear, cold Alaska night. The red glow of Simeon's cigarette lit the room in shadows and highlights that leapt and wove, dancing to the beat of his stories as the cigarette bounced in his lips. His soft, even voice transported us to another time, when the Sugpiaq people ranged from the Aleutian Islands in the west, east past Cordova, and north as far as Inuit territory. The territories of many peoples border on Prince William Sound, which is a great mixing ground where the history of various groups are mingled and confused.

In those days, "there used to be more witches." Simeon's eyes grew distant. He told us that witches could be seen at night, as recently as when Simeon was a boy, traveling overhead encased in a small fire, headed to another village to cast a spell. People do not use witchcraft so much anymore, though, since the introduction of Russian Orthodoxy, which is a stronger power. Simeon told us how his father, who was a reader for the church and the choirmaster, had succumbed to a spell and became very ill.

"My father lost weight until he couldn't get out of bed. Out the window from his bed, he saw the man who cursed him pass by, and he said to Mother, 'Call in the witch. I want to speak to him.' Then my father said to him, 'Look what you have done. I am nothing but skin and bone. But wait 'til you see what happens to you when I ask God to help me with you.'"

At this, the witch was frightened.

"Wait," pleaded the man. "Tomorrow I will bring you something."

The next day, the man returned to Simeon's house, bringing with him holy water and holy bread in a jar, which he mixed, and Simeon's father drank these down. He immediately vomited up a small octopus. The octopus had been in his stomach, eating all the food that came down. Had the octopus stayed in his stomach, Simeon's dad would have continued to waste away. Simeon's father chopped the octopus into small pieces on cardboard, using a knife dipped in holy water. He sealed the pieces into a milk can, which he pounded flat and threw into the bay. A day later, the man who had cursed him became very ill, as always happens when a witch loses his power.

"Octopus are also called devilfish," Simeon informed us. "I asked my mother about this once, but she said there was nothing to worry about. She said, 'There's no devil in the octopus, else we wouldn't eat them.'" Simeon laughed.

At the end of the evening, Tania and I walked back to our lodgings under a brilliant winter sky, so clear and hard with frost that the sky felt close and the stars swarmed around us like fireflies. We looked for the comet, Hyakutake, that was supposed to be visible at that time, but could not find it. I would find it a few nights later as I stood in the snow relieving myself at 4 a.m. Our lodgings lacked plumbing, and my bladder forced me outside in the early hours where I stood in awe at the silent radiance of the Alaskan night sky in winter. The comet stretched across the sky in milky-white splendor, its tail like a laser beam through the heavens. Straight and true, it pierced the heart of the Big Dipper and ran on into constellations unknown before fading out near the western horizon.

Simeon's tale highlights the common wild diet of octopuses and of his people. An octopus midden displays a smorgasbord of seafood remains similar in kind to those that humans also love. The accumulated midden of clam and crab remains reflect our shared preferences and reveal the octopus's large appetite.

The big clams are a favorite food of two other shallow-water predators besides people and octopuses. The commercial fur trade once endangered sea otters (*Enhydra lutris*), but now they are common in Southcentral and western Alaska waters. These energetic hunters dive to the bottom to capture clams. They may dig down into the seabed to catch their burrowing prey. Sea otters may use their teeth to crack open the clams to get at the tasty meal inside. Tooth wear, however, is a strong indicator of age in sea otters and likely limits their life spans. Many otters also use a simple tool—they find a good rock. Floating on the water, the otter places the rock on its chest and pounds the clam against it to break the shell. An otter may tuck the rock under its foreleg while foraging and use the same tool to open clam after clam.

The sunflower star (*Pycnopodia helianthoides*) is a giant of the Alaskan seas with an appetite for clams. This behemoth of ocean floor habitats has up to two dozen arms and grows to more than an arm's length across and over ten pounds. It is also lightning fast—for an invertebrate—and can outrun most creeping seafloor animals, such as snails, abalone, and nudibranchs. Scallops and the anemone *Stomphia didemon*, as well as other usually sedentary prey, swim away when this sea star approaches or touches them. Beginning in 2014, populations of these flowers and other stars in Alaska waters suffered the ravages of sea star wasting disease, exacerbated or caused by changes in the ocean linked with record-high temperatures of the previous decade. Along our shores, sea stars now show uncertain signs of recovery.

In parts of its range, the main diet of the sunflower star is the butter clam. Some prey, such as sea urchins, live on the surface of the seabed and are swallowed whole by the sunflower star. But to get at the burrowing butter clam, the star excavates gravel piece by piece using the many thousands of tube feet on the underside of its arms. The result is a large pit with sloping sides, as wide as the sea star and about as deep as my calf-high boots, from the bottom of which the star pulls a clam.

To consume the clam, the star grasps it with its tube feet. To attach, cells in the disk at the tip of each tube foot secrete a bio-glue that adheres to the clamshell surface. The star tightens its pull, and its muscles have a ratcheting mechanism that allows it to sustain the pull without further effort. The clam muscles gradually become weary. When the clam opens, the star everts its stomach into the gap, releases digestive juices, and absorbs the nutrients of the externally digested clam meat. Different cells at the tip of the tube feet secrete a de-adhesive that allows the star to release its grip when ready.

Sea otters and sunflower stars both leave pits with high berms on the edges, and they both leave behind empty clamshells. The otters take the clams to the surface and break them before consuming them. The discarded shells seldom land back in the pit where the clam was taken. The sunflower stars often consume them in the pit. The shells are not broken, and the two halves usually remain attached at the hinge.

What about the octopus? How did it get into this armored prey? No doubt, the octopus had tried simply pulling on the large butter clam shell to pry apart the two halves. Whereas the sunflower star has thousands of tube feet that adhere chemically, octopuses have hundreds of suckers in two rows along each of their eight arms. The suckers are small near the mouth, becoming larger moving outward until they become smaller again toward the arm tips. Arms of the giant Pacific octopus bear about 115 sucker pairs or 230 suckers per arm, which works out to just under 2,000 suckers total. The male's third right arm, modified for reproduction, bears fewer suckers.

The suckers can adhere with surprising force. A tenacious grip from even one sucker will raise a small welt on a forearm. Each sucker has a two-chamber anatomy, more complex than a simple dome suction cup like you might buy in a store. These complexities allow the muscular action of the suction cup to draw a strong vacuum on the water inside the cup, taking advantage of not only octopus strength but also the capillary forces that act on water in small spaces, and of hairlike microstructure surfaces that seal the inner chamber underwater. The pressure difference between that vacuum inside and the water pressure outside holds the sucker rim to the surface.

In shallow water, the limitation of sucker holding force is the cavitation strength of the water itself. That is, the water inside the sucker behaves a bit like a solid, resisting any expansion in volume until an expansion force is so great that the water cavitates—microscopic bubbles grow rapidly under negative pressure that finally allow the volume inside the suction cup to increase. At less than that high level of force, the suction cup holds tight. The octopus does not have to continue to work at holding on: the water tension inside the sucker also maintains the contraction of the sucker anatomy itself even when the octopus relaxes. This is how the octopus can cling to a surface while relaxed or asleep.

Octopus suckers adhere to anything they touch. I once gave Obi, a young day octopus, a toy to explore in her aquarium. The toy was a Mr. Potato Head, with pink ears, big red nose, eyes, and other body part pieces plugged in. I put a piece of shrimp inside the toy through the storage hatch in the potato's backside. The plastic potato was as large as the octopus.

Intrigued but cautious, Obi slowly stretched out an arm. Mr. Potato Head was not well weighted and barely rested on the bottom. A sucker or two adhered to the smooth plastic surface of the toy. On contact, Obi pulled the arm back, startled. But the sucker didn't release, and so Mr. Potato Head approached. Obi, now detached, leapt back in alarm. Without the arm pulling on it, Mr. Potato Head drifted to a stop.

The sequence repeated, startling the hesitant octopus a second time. Still, this large and vaguely aggressive object commanded Obi's attention. On the third exploration, Obi approached in dramatic guise with her mantle darkly veined against a white background, and her arms black except for contrasting rows of white spots. With her rear four arms on the rock and substrate, she raised her front four arms above her eyes, web splayed wide along the arm edges. Mr. Potato Head missed the drama of this moment, as it had drifted face down and was staring into the sand. With her arms spread this way, lined with adhering suckers ready to resist any escape attempt, Obi's gape was larger than Mr. Potato Head—larger indeed than Obi herself. The arms, surrounding the mouth as they do, extend as prehensile lips as much as they do as limbs, and they were now poised to engulf Mr. Potato Head.

On first contact, however, again Obi's suckers adhered, and this large but light object approached Obi, seemingly of its own volition. Obi's nerve failed and she retreated nearer her lair. Mr. Potato Head drifted closer before coming to a stop, and she eyed him warily, having yet to discover that his various parts pulled out in tempting ways, and that there was indeed a morsel to eat inside his hard but oddly light-weight shell. Obi was still learning with her suckered grasp about light toys in captivity; she would later enjoy engulfing Mr. Potato Head in her web and disassembling his parts.

THE GIANT PACIFIC OCTOPUS of Port Graham that had defeated the butter clam mentioned earlier had taken the heavy bivalve in its suck-ered arms. But she was to learn the limits of her strength. With many suckers on multiple arms attached around both halves of the clamshell, the octopus no doubt had tried to pull open this armored prey, perhaps with some patience. The sucker attachment requires no persistent force, but the octopus has to pull continuously with its arms to pry apart the clam halves, while the clam resists, applying opposing force to hold itself closed.

Watching an octopus apply this pressure can be a quiet affair. If the clam wins, this may be a motionless tug-of-war. With small clams, a giant Pacific octopus quickly and smoothly pulls the clamshell open, some-times with enough force to break one of the clamshells. The day octopus opens clams and even pries loose prey from their holdfasts on the reef with a series of sharp pulls, its whole body jerking until, sometimes with an audible crack, something breaks and the inner flesh is accessible. The day octopus is very fast, often breaking into its prey or moving on within less than a minute, whereas the giant Pacific octopus may spend a few minutes to tens of minutes opening prey by pulling.

In its battle with this large butter clam, the octopus tired first. This clam was too strong, and did not yield to the octopus's strength. So the octopus tried something else. On the outside of the clamshell were no fewer than five separate marks, two marks on one half of the clamshell and three marks on the other side. The marks were small ovals made by the giant Pacific octopus. Each of these was an attempt by the octopus to get through the shell.

These tiny oval perforations are drill marks. Octopuses have inside their mouths a radula, which is a rasping organ used to break up food. The radula apparatus is a distinctive adaptation found only in the mollusks. The radula itself is a ribbonlike membrane that runs between two muscle groups and lies over and between the bolsters, two muscular hydrostats. Our human tongue, as well as octopus arms and the elephant's trunk, are muscular hydrostats—anatomical constructions that use fluid pressure generated by muscle contraction rather than a rigid skeleton to allow movement. Inside the octopus mouth, the bolsters can direct the pressure of a bend in the radula ribbon. Along the length of the radula are rows of micro-teeth. Muscles at either end pull the radula back and forth, rasping it over and wearing away the surface against which it is applied.

The radula begins the work of drilling, and is sufficient itself to make a hole. But it cannot penetrate too deeply. Beyond that, the salivary papillae, also tipped with a few rasping teeth, must take over. The salivary glands secrete enzymes that break down the shell of prey such as crabs or this clam. The papillae delivers the caustic secretions directly to the drill site, chemically dissolving the shell, and making it easier for the tooth-tipped salivary papillae to rasp away. Toward the bottom, each tiny oval excavation, begun as a shallow depression by the rasping radula, narrows further to no more than the width of this slender salivary papillae.

When the shell is finally penetrated, the saliva is secreted through the hole, where chitinases and toxins begin to loosen muscle from its attachment and subdue the prey. Octopuses also drill many crab shells. The giant Pacific octopus in Alaska commonly drills crabs on the posterior portion of the carapace or on the claws themselves. The curled octopus (*Eledone cirrhosa*), which lives in the northeast Atlantic from the Mediterranean to Scotland, drills in this way through the eye of crabs, penetrating the cornea, to inject their digestive secretions. The eye is a particularly vulnerable spot on these armored crustaceans, and this tactic allows the octopus to immobilize its prey within minutes of capture.

On the large butter clam, not one of the five drill marks penetrated the shell. None made it through to the interior. The octopus spent many frustrating hours in its den eroding these holes into the shell but always tired too soon, or found that not only could the radula not reach deep enough, neither could the salivary papillae. The wearied octopus started over each time in another spot. That shell was thick!

Five attempts to drill into the armor failed, yet the octopus still was not defeated. On the rim of the clam, where the two halves met opposite the hinge, I found small marks. The mark on the left half lined up with the mark on the right, leaving a tiny gap to the interior of the shell. The mouth of the octopus contains a black and chitinous beak, reminiscent of a parrot's, but more fully surrounded by muscle. While squid beaks are often very sharp, useful for tearing into the soft flesh of the fish that comprise their diet, those of octopuses are often worn and blunt from applying force to hard shells. After a long battle using suckers and arms, radula and saliva, the octopus had beaten the clam by biting with its beak, and was finally able to chip the edge of the shell at its thinnest point. Even though the chip was small, it exposed the inner clam flesh. This allowed the octopus to inject saliva that paralyzed the clam and weakened the connections between shell and muscle.

Many prey do not present the challenges of this heavy butter clam. We find drilled clams and crabs quite often, without marks of multiple failed attempts. The shells of the large helmet crabs are lightweight compared to other crabs, and in this same midden were numerous helmet crab legs. Octopuses do not trouble to drill these crabs through the carapace. Instead, they bite through a leg to open them, leaving a telltale mark, or they simply rip open the crab with arms and suckers. Interestingly, young octopuses used to eating crabs subdue their first Pacific littleneck clams by drilling, but later they learn to pull these open, which they do in seconds.

Octopuses are well-equipped hunters, with a formidable tool kit of body parts to defeat the heavy armor of their prey—the radula and salivary papillae for drilling through shells, the beak for chipping shells, and the force of the suckers for pulling apart prey. This flexibility of methods is characteristic of the octopus, possessed of curiosity to investigate, a fearsome patience in persisting to the end, together with the will to abandon one approach for another and another, until the meal is won.

However octopuses are not only the hunter. They are also the hunted.

# *Watching Octopuses*

# 10

## Storied Octopuses

"In the story," I said, glancing up with trepidation, "there is a hunter."

Dr. Apela Colorado was listening. Apela is of Oneida-Gaul ancestry, and she devotes her energy to promoting Indigenous science, an ancient way of knowing used by tribal people for generations. Indigenous science seeks not only to understand but also to respect people and the natural world. Apela was about five foot two, with a broad face and luxurious thick black hair streaked with gray.

How was it that I was telling an Indigenous elder a traditional Aleut story? If she knew this tale at all, she knew more about it than I did.

She and I were in Anchorage at the annual workshop of the *Exxon Valdez* Oil Spill Trustee Council on the traditional lands of the Dena'ina people. That year the workshop emphasized the importance of Traditional Ecological Knowledge to understand natural systems. Apela was there to bring native voices and Indigenous science forward in the restoration process for Trustee Council support. She had just approached me to talk. Later she would say that it was my obvious passion for my work and genuine concern for octopuses that drew her to me.

I told Apela about the work I was doing with octopuses, and what I thought it might offer for the work she does with traditional knowledge. In allegory, predators and prey are enemies, as was the case in the story I was about to tell Apela. Octopuses are hunters, but they are also soft vulnerable prey, and the story I wanted to share with Apela intrigued me. Given a history of conflict between researchers and Native communities, it piqued her interest, too. Maybe there was something for us both to learn.

"So, what is your next step?" she asked. I had just finished the fieldwork, and I expected to submit a final report to the Trustee Council. However, I was toying with a few ideas to continue working with octopuses.

"Tania and I have found or been told octopus stories from coastal northwestern natives—the Aleut, Alutiiq, Eyak, the Tlingit, and Haida. I want to put them together with experiences in the field and use the framework to write about octopuses."

I showed her a drawing illustrating one of the legends, and a photographic collage of octopuses. Apela seemed fascinated by a deep ocean image of a giant Pacific octopus.

"What is the story?" she commanded, looking directly at me.

"In the story," I said, drawing a deep breath, "there is a hunter.

"He hunts sea otters with the other men in his village. However, when he goes out hunting, he never gets a single otter. The other men in his village are successful, but this man, he never gets any otters. One day after a hunt, he is paddling back to the village, again without any otters for his troubles, and thinking about what he should do to be successful. A great old octopus swims up alongside him. The octopus is as long as his baidarka. It grabs him and takes him below the sea to the octopus's house under a boulder.

"There this huge octopus says to him, 'Why do you never get any otters when you are out hunting.'

"'I don't know,' replies the man. 'Whenever I go out, I never get any otters.'

"'Apparently, you have been eating in the evenings before you hunt,' says the great octopus, and he reaches with one arm into the man's mouth, and down into his stomach. He withdraws his arm, clutching grass and weeds.

"'Apparently, you have been walking in the evenings before you hunt,' the octopus continues, and he rubs the bottoms of the hunter's feet, scraping off the skin, the calluses. He does the same to the hands.

"'Now when you hunt,' the octopus says, 'you will have great success. After each success, be sure to lower to me the white heads of the sea otters.'

"To this, the hunter readily agrees, and the octopus takes him back up through the sea again, so that the man may continue on his journey home.

"On the next hunt, the man cannot miss. Every time he throws his spear, he takes another otter. On the way home after the hunt, he lowers to the octopus the white heads of the otters he has killed. This continues for some time, until one day the hunter has a desire for a white sea otter head of his own. He keeps one and lowers the head of a black sea otter to the octopus instead before returning to his village.

"Sometime after arriving there, he feels the ground shaking. Then he hears talk that the bay beyond the village is turning red. The hunter hears that an octopus's arms appeared above the water in the middle of the bay. At this, the man realizes that it is because of him that this is happening. The octopus throws the sea against the village and all are swept away, it is said."

"Do you know what the story means?" Apela asked as I finished relating the tale.

I puzzled for a moment over this. The story was partly about the hunting taboos of the Aleuts and proper behavior. Another level in the story, on which the elder octopus in the story spoke to the ecologist in me, is that octopuses gave aid to the enemies of sea otters. I had seen otters in Prince William Sound and Cook Inlet floating on their backs, chewing on large red octopuses draped over their chests and bellies. I had found remains dropped from such meals during the intertidal surveys. Was the octopus speaking to me about its predators? The enemy of my enemy is my friend? These interpretations were not what Apela meant, however.

"Just like Western science, Indigenous science has rules," Apela said as she tapped the drawing. "And good things can happen to those who respect them. But if the rules are broken . . ." Her glance fell meaningfully to the last panel of the illustration, where the octopus swept up the sea into a tsunami that hung above the doomed village.

I asked a question that had been on my mind. "Do you think it is appropriate for someone like me to use these kinds of stories in my own work?"

She looked at me, understanding what I was asking. The stories belonged to the cultures that told them, no matter who wrote them down in the books where they appeared.

"The giant octopus is a central, sacred archetype for Indigenous people," she said. "It symbolizes and is a key to unlock the knowledge you are searching for. Be careful. You must ask permission, especially for the stories that have not been published."

THE STORY OF THE ALEUT HUNTER and the huge octopus, concerned that the hunter wasn't killing enough predators, came back to me one day near Busby Island in Prince William Sound. I was underwater, holding a yellow mesh goody bag containing an octopus ready to be released. With typical caution, the octopus extended two slippery arms slowly downward, interested in the rocks and open water in view outside the mouth of the mesh bag. The octopus was hesitant to move from the bag to a new place where dangers might lurk.

As I held the bag, hovering over the bottom and waiting for the octopus to exit, my dive partner tugged at my arm, gesturing. I didn't see what she was pointing to, but clearly she didn't want me to release the octopus at this spot. I looked around again through the shallow green water—nothing. My dive buddy shrugged.

We moved a small distance to where the edge of the kelp stands met the sand. At the base of a rock, I began again. Another tug on my arm— no, no! She pointed ahead, and we moved farther until we were directly above a thick understory of broad brown kelp, spreading into an unbroken canopy of blades just four to six feet over the bottom. I had tried swimming below these canopies, but ropelike kelp stipes that attached the floating canopy with holdfasts to the rocks below allowed only the slightest progress for a diver in scuba gear. Here, my dive buddy nodded. I began a third time to release the octopus.

Still unsure what had bothered my buddy, I was now more vigilant. She was the same. We faced one another, each peering over the other's shoulders into the retreating murk, my buddy alert for whatever had aborted both earlier attempts, me looking for anything that could pose a problem—stinging jellies, tangling lines, drifting dangers. I saw nothing. Facing the bottom, I opened the bag that contained the octopus. Again, I sensed my dive buddy approach above me as shadows changed the quality of light where I focused on the octopus, my bag, and the kelp. I looked up to see if she was waving me off yet again.

Instead, just inches from my mask and descending vertically from above me was the large head of a curious juvenile sea lion, intent on the octopus meal soon to be released from the yellow goody bag. For a brief moment I looked into her big round brown eye, noted the tiny ear flap, the black nose with tightly closed nostrils above stiff whiskers, and the bone white base of curving canines. Noticing my awareness of her, with an easy flick of her flipper, she turned and was gone. Now I understood what my partner had seen. Our own clumsy diving activity had drawn the attention of a curious and hungry predator. Were we to release our octopus here, she would have little chance to become invisible to the sea lion. We ended the dive, deciding to release her the next day after the sea lion had left the area.

The risk of meeting these predators shapes the octopuses' world. Their first line of defense against predators is avoidance. Octopuses hide in their dens, crawl quietly through dense cover in search of their meals.

They choose the largest crabs to keep their foraging brief, and perhaps shift the time of day when they are active to when their predators are not. They often return to the den, or another safe and sheltered spot, where they open and consume their prey. They keep themselves out of sight.

There are important tasks out of the den to accomplish, though. Food and perhaps mates must be found. So the sensory systems of their predators have shaped how octopuses avoid being found.

The sea otter and the sea lion are both mammalian predators. Mammals on land can use smells carried on the air to locate or track food. A few mammals, including otters that eat octopuses, do use smell underwater. A Eurasian otter (*Lutra lutra*) was filmed exhaling small bubbles from its nose and then sniffing them back in, the odorant molecules from the water now dissolved in the bubble membrane and humidity inside it. The otter found a dead fish underwater in the dark using the bubble method. Moles and water shrews use the same underwater sniffing method when foraging.

This Eurasian otter species does not occur in Alaska—our two otters are the North American river otter (*Lontra canadensis*) and the sea otters. It seems likely these may use the bubble method to smell underwater, although no one has yet demonstrated this for either species. Fish, rays, and sharks also may smell out their prey in part by the chemical trails they leave.

In the tropics, day octopuses are circumspect, and moray eels may pose enough risk to drive tidiness. In these waters, day octopuses eat away from the den or carry remains of their meal some distance from the den before discarding them. Perhaps this makes their odors more difficult to detect. The moray eel predators (unlike sea otters in Alaska) are well suited to enter an octopus den, if they can find it, and drag out an octopus. Moray eels of many different species are a constant danger to octopuses in the tropics. Although morays as a group occur worldwide, they reach highest diversity and abundance on tropical coral reefs. Moray species are nocturnal and rely on scent to find their prey (although a few species forage visually).

Sea lions, however, might not smell underwater. As I saw, their nostrils are closed when diving. Olfactory hunters may be scarce in Alaskan waters, and giant Pacific octopuses are messy housekeepers. They leave a conspicuous midden of prey remains just outside the den.

Mammals that become aquatic usually lose some sense of smell, or fail to transfer it to the water. Most species of whales and dolphins have lost their sense of smell and no longer develop olfactory parts of the nervous system. As they became aquatic, animals such as snakes also lost some, but not all, of their sense of smell—sea snakes, which are fully aquatic species, more so than amphibious snakes.

With the sense of smell absent or diminished, these marine predators must rely largely on other senses. Hearing, touch, and vision are important. Whales and dolphins rely on sound, hearing, and echolocation. Seals and sea lions can track the water vortices left by swimming fish using very sensitive whiskers. Manatees and dugongs—which are grazers not hunters—also use very sensitive whiskers to forage. Fish have a similar ability to sense with their lateral lines, and cephalopods have lateral line analogs with pressure-sensitive cells that sense movement in the water.

Although the rattle of our scuba breathing no doubt was audible, the sea lion may have found my dive buddy and me underwater using their vibrissae, following the vortices of our kicking fins. Years earlier during my first underwater encounter with a sea lion, I was swimming along above the kelp through murky water. My dive buddy followed closely. We had to be close to stay in sight of each other.

I felt a tug on my fin, our agreed signal for communication. I turned and looked at my buddy, who made an uninterpretable gesture, and pointed into the murky water. She shrugged. I returned to swimming. Another tug; more uncertain sign language from my dive partner. On the third tug, I turned back to her quite sharply. She gestured to the distance, pivoting as she pointed, her finger following an arc. Out in the indicated direction at the edges of visibility, I saw a dark torpedo silhouette against the lighter water making a fast circle around us. A sea lion.

The first I had seen underwater. I was thrilled.

I turned back; my companion's eyes were wide, watching me. I felt through my thick neoprene dive hood a pressure above the ears on either side of my head. For a beat I was unsure what was happening. Then I was pulled upward twice. The incident lasted only a moment, but I was relieved when the unseen hold on me vanished.

My dive buddy later recounted how the sea lion had been intrigued by my fins, grabbed them in its mouth and tugged. We had no pre-arranged sign language for this, or even for *sea lion*. When I stopped kicking to peer at my dive buddy, the sea lion had swum a wide arc around me and then took my head in its mouth, and had an experimental tug to learn more about what I was. We encountered the sea lion unexpectedly—but when planning to capture sea lions to attach tracking tags, working scuba divers wear bicycle helmets underwater for protection from just such ever-present curiosity.

Crawling octopuses, however, may not leave swirling vortices behind them in the water. The pinnipeds (seals, sea lions, and walruses) and sea otters, as well as fishes, also have many adaptations for acute vision underwater. They rely on sight to find their prey.

In shallow tropical seas, the sun is bright. The water is low in nutrients and often clear. Fish are plentiful and abundantly colored, and clearly also use acute vision to hunt. Vulnerable octopuses would be wise to hide, and to come out only at night. Yet this is the home of the day octopus, named because it is active during daylight. Day octopuses stay safe in a protective den at night. This behavior avoids nocturnal predators such as morays that hunt by scent. However, day-active octopuses are then at risk from daytime predators hunting by sight.

Visually oriented predators active during the day include seals and sea lions, otters and birds, and many species of fishes. Hiding in dens and under cover is the octopuses' first line of defense against visual predation. When they must go out, however, the second line of defense octopuses use is hiding in plain view. Their strategy is to look like something, just not like an easy meal. The octopus stands a chance among these visual predators, because the octopus is a master of disguise.

Their first disguise is body color, with two ways to hide the body outline. Disruptive coloration displays large blocks, stripes, or bands of dark and light. Through even a modest distance of water, the predator eye sees these dark and light blocks as different objects. The outline of an octopus as an easy meal is lost. A mottled display is a pattern of about the same brightness and element size as the background where the octopus sits. A good example is a bed of gravel. The varied texture of the gravel and the varied darkness of an octopus's mottle display combine to obscure the outlines of the octopus. The octopus vanishes against the background.

The second disguise is body texture. Octopus skin in a relaxed state can be smooth. But in disguise, octopuses can raise bumps, tessellations, folds, or papillae on their skin. The folds along the mantle of the giant Pacific octopus wave and sway like kelp fronds, and make the animal difficult to see. The small pointed papillae of the red octopus resemble the texture and color of small coralline algae. The day octopus raises large broad papillae crested with pale colors, like digitate small-polyp coral tips catching the shallow tropical sun. From a meter or so distant in the dappled glinting light of coral seas, the reef and the octopus sitting atop are indistinguishable.

The day octopus does more than match skin texture to the coral. The third disguise is composed of postures and movements. Tucking its arms under, with eyes raised and papillae erect, the octopus is disguised by posture and shape as a conspicuous coral head or an algae stalk like those surrounding it.

To do this, the octopus must pick from nearby objects a specific thing to imitate. In the nearshore waters of Mo'orea, day octopuses might imitate spiky-bladed, brown-stalked *Turbinaria ornata*, the most abundant brown alga on the disturbed coral reefs of French Polynesia. In coral rubble or on the less disturbed outer Great Barrier Reef at Heron Island, day octopuses might sit at the edge of coral outcrops, mantle pressed tight to the body, in the guise of a jutting coral head.

On a sandy flat in the shallow waters off Sulawesi, one of the main islands in Indonesia, the algae octopus (*Abdopus aculeatus*) might erect columnar branched papillae at the tip of its mantle and hold its arms up in the water in stiff curved or coiled positions to resemble a tuft of marine algae.

Disguising movements is difficult. Visual animals track moving things more easily than still ones. To the predator eye, nothing is more conspicuous than motion. The octopus takes advantage of this, and tries to look like something other than an easy meal. She jets up into the water, turning uniformly dark in color as she starts to move, and briefly swims among a school of passing reef fish, conspicuous but not as an octopus. Returning to the bottom, this odd fish among the swimming school vanishes when her movement stops. The octopus in an instant adapts the color and texture of a chosen object from the surroundings.

In motion, the octopus web and arms spread wide and billow to pounce on a crab or other prey at the bottom. Such a conspicuous capture cannot be hidden from an alert predator. But maybe the octopus can become unrecognizable—not an octopus, not a meal. The wide web and arms blanch soft white. The octopus displays a bold white stripe between the eyes extending backward, visually separating the two dark or mottled sides of the head and mantle. The web appears like a white handkerchief in the water. The mantle may appear as something separate or as parts of the background. As the pounce is completed, however, and the prey ensnared, movement stops. Again the mottle camouflage reappears. The stark white target vanishes from a predator's field of view the moment the octopus movement is complete.

As they move in broad daylight across the reef, these octopuses change their appearance repeatedly. When stationary, they match a nearby object. Momentarily in motion, they become white or black, as visible as possible. In this moment, they lose themselves among nearby fish in motion.

The day octopus in slightly murky water will pull all her arms beneath her. Pressing the mantle tight down on the body, she assumes a humped shape like a worn coral head or a rock. In this guise, she creeps cautiously across the open. Her pace is slower than the movement of algae or loose debris swaying with the waves above. The day octopus appears not to move, but she covers the open distance.

The algae octopus, arms aloft like a tangle of algal holdfasts, walks briskly on two legs across open ground. The mimic octopus (*Thaumoctopus mimicus*) when moving assumes the shape of dangerous or inedible fish. It holds all arms in a plane against a flattened mantle. The front arms are curved in broad arcs backward. The mimic octopus becomes like a flatfish with transverse stripes across its arms and mantle. With arms banded black and held at jutting angles to the sides, it resembles a lionfish. With rear arms in a burrow and front arms held out left and right in waved arcs, it becomes a sea snake resting on the bottom.

Each octopus has many choices moment to moment. Moving across the reef flats, the day octopus changes on average three times per minute when mobile. Appearances include mottles or disruptive body patterns and textures as they forage slowly across the reef but also bold disruptive patterns so as to appear little like an octopus. Faster movement includes brief mimicry of fish. Overall, octopuses adopt more than a dozen different appearances while out of their dens.

Predators must have a search image when looking for meals. Detection of vulnerable prey drops when the prey doesn't match the search image. Even three choices can cut the number of detections in half. This is why simple disguises work, if not perfectly, at least most of the time. Changing disguises a few times per minute, octopuses avoid being seen, recognized, and tracked, even while moving about in broad daylight.

OCTOPUSES LIVE IN HOSTILE SEAS, full of watching predatory eyes in daylight, and listening, sniffing, and feeling predators at night. Octopuses are adept at forestalling notice and avoid the need to fight with their hungry enemies; enemies that commonly swim clueless over a motionless octopus, noticing nothing.

Out of sight or in plain sight or in disguise, when hiding fails, however, a predator can focus its attention on its octopus prey.

Even then, octopuses are formidable opponents.

# 11

..............

# Octopus Adept

**Underwater off the Big Island, Hawaii**

"Come. Come."

The dive master's signal from thirty feet away in the clear warm water interrupted my reverie. I glanced down in front of me where a small charming day octopus, awash in splotches of warm umber, cream, and translucent red cacao, was watching me, motionless. She'd been still for a while, a few cautious right arms stretched along a crevice in the reef, one bold left arm extended out toward the open, the rest curled. I glanced up at the dive master again, now a few feet closer, his sign language more exaggerated.

"Come!"

He wanted me closer to the group that, along with my dive buddy, was another fifty feet beyond him. The water clarity in January off Hawaii was disorienting after diving so often in the emerald murk of glacial Prince William Sound.

At my slight movement, the octopus retreated. I was making her nervous, and I was making the dive master nervous. Heading toward him, I left her. Immediately, the diver master turned, pushing through the water, now and then with a glance to ensure I was following. Distances were deceptive—I was longer reaching him than I expected, and by the time I arrived, he had in his hands another squirming octopus. New to Hawaii and to tropical diving, I had told our dive master that I was hoping to see octopuses. He had pulled me from one to see another.

This one was battered. Three arms remained intact—the rest existed as stumps partially healed and beginning to regrow, or scars and shreds notched where the other arms belonged. The dive master proudly held this shattered animal out to me, pleased with his find. I sighed. I had been called from one octopus exhibiting its natural caution to this one, captured in the diver's grip. I would have loved to see this animal in place before it had been disturbed. I watched a moment as she snaked an arm between the dive master's fingers and along his wrist, seeking her escape.

I motioned to the dive master: "Put her down, return her to the reef."

He nodded, swam forward a bit, and tossed her to a hole in the reef into which she eagerly fled. The dive master turned to follow the rest of the group ahead. I lingered a moment to watch the octopus settle, when from the hole into which she'd escaped emerged a moray eel, the tripedal octopus pinned in its happy rictus around the unexpected gift. The dive master had accidentally delivered her to the burrow of perhaps the very predator to which she had previously surrendered arms.

The life of this day octopus had been one of constant danger. When she failed to escape notice, the soft-bodied octopus was vulnerable. Given half a chance, she employed an impressive range of feints, deception, and sacrifices to try to escape with her life. Her successes in escaping by the narrowest of margins were written on her body. I saw this also with many others octopuses—scars and missing limbs or parts of limbs are common.

Her first lines of defense—hiding and camouflage—had at times failed this little octopus. I imagined in those moments she used displays so drastic and dramatic that her disconcerted predator hesitated, winning her the chance to flee.

The Greek god Ares the destroyer, signifying the brutality and destruction of war, rode with his son Deimos, the personal god of dread. Together they sowed terror and disorder in battle. When spotted and threatened, the octopus, like Deimos, attempted to evoke terror: her arms—before she lost them—curled outward, webbing spread wide and flat, the visible edges of her suckers black. As this large round surface area blanched white, black clouds coalesced in a wide ring outside of the eye. In the center of this dark dreadful storm, the animal's eye was encircled in a white ring.

Visually the entire octopus transformed herself into the face of some stark terror—a deimatic display. The white pupils of her skin-painted eyes were three times the width of the actual eye, and the apparent eyes covered her entire head. Her face white and black, she made a menacing feint, as if to engulf her would-be predator. All of this happened, I imagined, perhaps only once but maybe many times, each in an instant. The attacker saw in that moment the error it made—and the risk that awaited! The small target was revealed to be the head of a previously unsuspected giant. In this blink of the predator's second glance, the octopus changed again, pushed into action, left an ink cloud behind to hang just where she used to be, and jetted away.

She ejected her ink, when she used it, from the ink sac through the same siphon that focused the jet-propelled escape. Most shallow water octopus species have ink, like the day octopus that is active in a sunlit visual environment. Deep-sea species of octopuses often have lost their ink sac, living as they do in dark environments.

Octopus ink is composed of melanin and mucus. The ink gland secretes the melanin-rich black ink into the sac where it is stored. Cephalopod ink gave name to the color sepia (the word originating from the Latin *sepia* "cuttlefish," the ink of the cuttlefish and octopuses being the dark brown pigment in question). Sepia ink is long-lasting when used in writing and art. The dark brown to black colors in octopus skin patterns come from chromatophores that contain melanin. This dark ink may have evolved just after chromatophore melanin in the skin or contemporaneous with it. The skin color-changing system also includes chromatophores with yellow to red pigments. Other parts of the skin produce a range of colors from white to intense blue using iridescent, reflective, and refractive structures.

The funnel organ, attached to the funnel itself (that is, the siphon), produces the mucus that the octopus adds to the melanin ink as it is ejected. The mucus is carbohydrate-bonded proteins suspended in water. Together, the mucus and melanin form a liquid sepia colloid with cohesion and slipperiness. The two parts, separately secreted and mixed in the moment, allow her to squirt ink of different forms. With less mucus, she ejects a diffuse smoke screen. With more mucus, the ink coheres in the water as a cloud, called a pseudomorph, about the same size as the fleeing octopus. It is likely that octopuses have a sense of how they deploy their ink, although we do not yet know this with certainty.

Squids that use ink may have a notion of the shapes they create. Deep-water squids in Monterey Canyon release ink when disturbed, doing so most commonly in the upper mesopelagic twilight waters (depths of about six hundred to three thousand feet). In this study, squid escaped rapidly concurrent with their release of pseudomorph ink that created dense coherent clouds (that might be mistaken for the escaping animal itself). By contrast, threatened squids remained close to other ink shapes they released, such as clouds, twisting ink ropes, or puffs. Squids used these as smoke screens, as defensive mimicry of elongated stinging siphonophores (midwater colonies in the same taxonomic order as Portuguese man o' war), or (within small squid groups) possibly emitted the puffs as warnings or to share some other message.

Let's return to the possible history of the unfortunate day octopus, just as she's escaping a moray eel. As the ink blob hung in the water, the predator of the little octopus at this point was certainly confounded. If it persisted in the attack, the predator's jaws would have closed around an insubstantial cloud where it expected prey. Nowhere nearby was any fleeing meal to pursue. The octopus now was distant, and she had again stopped, camouflaged, motionless, and indistinguishable from her surroundings.

Baby turtles will pursue an octopus, and when it inks and jets, the turtles will attack one of the pseudomorphs. However, having once snapped on an octopus ink blob, that turtle will not again attack an octopus nor bite into its ink cloud. Ink, it seems, tastes bad to some predators. However, while octopus ink deters the turtles, the smell attracts some moray eels. The inks of different species appear to have different flavors to different predators, including to humans. People find the ink of a squid to be harsh and pungent. But the mellow, velvety tasting ink of the cuttlefish is palatable to people. In fact, ink of the cuttlefish is used in Venetian black-dish cuisine. Before consuming a cuttlefish, on the other hand, dolphins are careful to release the ink by beating their cuttlefish prey against the substrate.

As for the little day octopus and the eel, the moray, perhaps, gliding sinuously around the corals in Hawaii one day, came upon the little octopus unexpectedly; or perhaps the fleeing octopus, having jetted away from one threat, inadvertently landed on another. Something like this must have happened to the little octopus in the months before she died, something similar to an attack recorded by snorkelers in Hanauma Bay. This time, the moray struck, surprising the little octopus. She blanched again around false eyes that darkened, but at the same moment she jetted away. Aiming for the octopus's eyes, the moray instead clamped its jaws high up on one muscular arm. The sinuous moray twisted and turned, thrashing the octopus in a cloud of silt. In a moment, caught and unable to flee, the octopus engulfed the head and jaws of the eel with her other arms and her web. Another arm, as the moray twisted, caught the tail. Now the small octopus and the larger moray were wrestling.

The octopus was ghostly white with a network of black blotches over her head and mantle. The moray clamped her arm in its sharp-toothed mouth; the octopus held the moray's mouth closed, preventing the moray from any other bites. In this hold, her arms were tense and tight, every sucker and muscle engaged in the headlock. More suckers along the arm holding the tail latched on, and that arm gradually encircled the body of the eel, holding its tail tight to its head.

The moray found itself with jaws held shut by the octopus arms around its mouth. It felt a suffocating octopus arm reaching inside its gills through the gill slit. Octopus suckers and flesh covered its eyes and nostrils. The octopus held the eel's tail, and so the moray was unable to swim. It thrashed.

But the moray was as agile as the octopus. In two or three twists of its tail, the moray detached the clinging octopus arm, which released reluctantly, almost sucker by sucker. With its tail free, the moray tried a new trick, one at which it was an expert. It curled its tail up over its body, and then around the far side and under again, and finally back through the curled loop above its body. A knot!

The moray can slide the knot forward against anything of which it has taken a bite—a larger fish, a carcass, or prey attached to something—and apply pressure to tear flesh. The overhand knot is a simple one, but morays tie themselves into at least five different types of knots, including two types previously not found in the most comprehensive knot-tying manuals.

Although it could not open its jaws nor clear its gills with the octopus clamped on and in its mouth, the moray slid its body through its own knot, moving the knot toward its head. The moray slipped this knot over its head with a sudden jerk. In doing so, it tore flesh from the octopus, entirely severing the one arm clamped in the moray's mouth. The jerk of the knot clearing the moray's head shoved away the rest of the octopus that was covering its eyes and nostrils and holding closed its jaws.

Finally free of the suckered menace clamped over its head, the moray lurched forward, straightening out its long body. The octopus fell to the sea floor behind it. The very moment she was free and hit the bottom, the octopus jetted herself up into the water, a trail of ink clouds spinning out in her turbulent wake. The moray turned at just this moment, and bit into the second ink cloud, but the octopus was no longer there. The entire battle lasted less than thirty seconds.

The eel regrouped, looking exhausted by the fight. Its sole prize, a portion of one arm torn off the damaged octopus, and partly in the eel's open mouth, still clung by suckers over its nostrils and its eyes, obscuring scent and vision. Even clear vision would not have altered the outcome by now, however. The little octopus, although damaged, having lost one of the several arms she would be missing before her final moray encounter, was no longer nearby, and no longer discernable against the background of the reef.

Octopuses have more than one way to leave an arm behind to escape predation. The arm will continue its hold after severing, and may also twist and turn. Long-armed octopuses of the genus *Abdopus* (including the algae octopus) have a zone of tissue weakness in the arm, where, by a combination of nervous signals and pulling, they are able to detach the arm, much as a lizard can detach its tail—a process called autotomy. The day octopus in Hawaii is a muscular species lacking this weak zone, but the sacrificed portion of the arm is still capable of clinging to its attacker, as it did to cover the eel's face. A predator with an arm to eat is likely to forgo further pursuit of the fleeing octopus.

Even in a nonautotomizing species of octopus, a single amputation is not as serious an injury as it may sound. Octopuses do not bleed their copper-based blood, which is pale blue when oxygenated but clear when deoxygenated, from an amputation wound. Their muscles constrict directly around an injury or amputation, pinching blood vessels and closing them off by muscular action. The muscular contraction of an uninfected wound plays an important role in healing, pulling skin down to reduce the size of the wound during the first few hours post-injury. A lost arm tip will heal and begin to regrow within a few days. Given time and good habitat, the octopus can regrow a lost arm.

The loss of an arm or arm tip is a constant risk given the exploratory lifestyle of the bottom-dwelling octopuses. Arm damage is common in wild populations.

So pervasive is the threat to octopus arms, that the arms are sensitive to light. When the eyes are in darkness but a bright light falls on the ends of the arms, the octopus will move the arm tips out of the light and into a dimmer area. This occurs even when the octopus cannot see her own arm tips in the light. She pulls them close, where they presumably are less visible and less vulnerable to nipping or tearing predators. This concern is severe for male octopuses, whose mating success depends on an intact third right arm modified for passing sperm to the female. Males take care for this reproductively critical arm. Even when out actively foraging or exploring, the male curls the third right arm under, close to the body, a behavior not seen in females.

A persistent predator, or a merely curious one, may try to get at an octopus even as she is secure in a den. Something like an eel may be able to follow deep into a den or crevice in the reef. Octopuses can block the den entrance to prevent this, sometimes as easily as by facing their suckers outward in a menacing wall. Depending on what is available nearby, the octopus may pull small stones around the den, hold a flat shell across the entrance as a barrier, or use other objects such as a marine sponge or even a discarded beer bottle as a barrier.

The coconut octopus (*Amphioctopus marginatus*) carries this defense a few steps further. These small octopuses live in open habitats almost devoid of cover. They make their dens in a clamshell or half of a coconut shell. Both halves of the clam or the coconut shell are better still, and the octopus will sit in one half, and pull the other half overhead, closing the door on its den. Good shelters are rare though, and the world is dangerous, so a coconut octopus with a good den will tuck both halves up under the arms, carrying the bulky shelter along while walking across the sea floor in a rolling gait on just two legs. This bulky apparition not only doesn't look much like an octopus, but if the octopus is menaced, the shelter can be immediately occupied for protection.

Octopuses have another defensive use of shells, seen in the 2020 Netflix film *My Octopus Teacher*. When pursued by a scent-tracking shark from shelter to kelp canopy, even briefly leaving the water over rocks before returning below, the film's octopus teacher exhausted all options of camouflage and inking escapes. With neither den entrance to block nor a coconut shell for protection, she attempted to fool the predator. She picked up dozens of shells and stones in every sucker, and wrapping her arms overhead with suckers outward, she became an immobile armored ball sitting on the open sea floor.

This did not fool the scent-tracking shark however. The shark grabbed her and thrashed the ball of octopus and shells. The octopus had to abandon her suit of mail to flee, but she was not yet quite done with the shark. Moments later, the octopus was on the shark's back, still holding a few shells and comfortably riding along behind the shark's head as it swam. No octopus was in its sight, and any octopus smell was somehow always behind the seeking predator. The shark passed close by some kelp and rock, and the octopus quietly departed, dropping her remaining shells. The shark was none the wiser despite being schooled by the octopus teacher.

High-stakes predator encounters are not rare for octopuses, who are masters not only of disguise but also of misdirection, feints and bluffs, and eventual escape. Not every octopus succeeds, however, but each successful maneuver buys the octopus more time before the next dangerous moment arrives.

# 3

## Reach

# Sensation and the
# Grasp of Octopuses

# 12

## Seeing Octopuses

**Underwater, South of Sydney, Australia**

I t is a languid midday pause sixty feet down on the seabed. The morning rush is over. Fortescues—small white and black Australian relatives of stonefish—dot an accumulated bed of discarded scallop shells. Banjo Sharks, with Cubist white and brown trails on their backs, overlap one another in placid repose. A gloomy octopus (*Octopus tetricus*) sits with its eyes just above the rim of his den, still except for slow ventilations.

The octopus's suckers under each arm line up in inverted orange ramparts, each sucker with a black rim. He displays a frontal white patch, and a pale oval just behind his eyes runs two-thirds of the length of the mantle.

Without clear cause, a resting scallop startles and jets itself up, clapping the beat of its own propulsion. The octopus turns his head with interest at the motion. One Banjo Shark swims into the scene to rest near another. The clapping stops and the scallop slews and twists downward through the water to rest again at the bottom. The octopus, attentive, stretches from its den toward the resting scallop, and in a moment extends two front arms in a pincer move to either side of the prey and curls one around the shell.

Both the scallop and the octopus are mollusks. Both see and taste, and both swim by jet propulsion. However, the scallop's vision and swimming method limit what it can do. The scallop may not see precisely where to go, and for many species of scallops, their erratic swimming cannot accurately direct them. In contrast, the octopus has eyed his prey and has now caught his lunch.

Using one arm, the octopus tucks his lunch under the web. Seeking the security of his den, the octopus makes a small jet back to home, and at this point drops the scallop again.

Why catch the scallop, his staple food at this location, but not eat it? After moving it under the web toward the mouth, the food is out of sight—leaving the visual sensory system. This was not the first round for this octopus and this scallop. About an hour earlier, the octopus had been out foraging, and captured the scallop. He brought it back to the den, dropped it, and seemingly forgot about it. The scallop sat for five minutes where it had been dropped. Then, up swam the scallop, and the octopus reached out toward it. The scallop landed just out of the octopus's sight that time, and was again ignored.

In these seemingly simple acts of collecting and then releasing prey, on display are notable features of sight, reach, and touch unique to octopuses. First, vision plays an important part in detecting prey, and the vision of predator and prey are each worth consideration. Second, the act of reaching could be surprisingly complicated for octopuses. Third, touch and taste are both important aspects of the octopus's world. The interplay of these systems of sensing may result in the undecided nature of this octopus's interaction with the scallop.

THE EYES OF OCTOPUSES are large, proportional with the role of vision to their perceptual worlds, and are widely spaced on top of the head, their dog-bone-shaped black pupils set horizontally across the white irises. They look out predominantly to each side. Like their skin, the color of the iris can change from white to golden-yellow or darker shades. Each eye sits in a large round mass of muscle at the top of the head. In a more attentive posture, the eyes either raise up slightly along with the head or pull down a bit. The raised eyes are expressive, like human eyebrows, or the ears of a puppy.

Octopuses are bilaterally symmetrical, and this is revealed also in the location of eyes to each side of the head and forward. Vertebrates have this left-right symmetry, too. Octopuses hold their eyes level, even when their body is not level or when in motion, maintaining a horizontal axis drawn from one eye to the other as if there were an internal gyroscope resisting other orientations. A complex vestibular system manages this, providing information about the direction of gravity as well as about octopus acceleration forward, backward, or pivoting.

Octopuses often move at a diagonal from straight forward or backward. This allows them to keep the approaching view centered on the most acute portion of their eye. The center band of the retina, parallel to the pupil, has the greatest visual acuity, much as people see the most from the center of our field of view. But because octopus eyes face mostly to the sides, in moving obliquely the octopus holds in view the direction of movement in the center of the fovea of the leading eye where acuity is highest. When moving straight forward or back, the landscape ahead will lie just on the edges of both left and right visual fields where vision is less keen.

There are two additional striking facts about the retina, relevant to the octopuses' visual world. First, octopuses do not see color. Second, they do see the polarization of light, which we do not see.

Blue wavelengths (around 430 to 475 nm) penetrate farther into the water than other colors of the visible spectrum, with red penetrating the least distance through seawater. As water depth increases in much of the ocean, blue light is the last to be available. The underwater world is thus often without color, except in the shallowest seas where some wavelengths are not fully absorbed by the overlying seawater. Even at the shallow depth of sixty feet, where our octopus has seen the scallop, photographs are devoid of reds and oranges without color correction or strobes generating new light underwater. When I dive, my mind here still perceives color, but that is partially dependent on knowing from the use of a flashlight or other sources, the color of the object when properly lit. I am fooled when I have no prior knowledge of whether a sea star is a dusky tan or vibrant red.

Primate perception of color uses three different receptors (trichromatic color vision), each with molecules sensitive to different wavelengths of light. Peak sensitivities are at about midnight blue, lawn green, and somewhere in the available shades of yellow, this last with a broad receptivity extending into the red end of the spectrum. The brain composes our sense of color from the relative intensity of response of all three receptors to the spectral distribution of the perceived light. Those with heritable red-green color blindness have deficiencies in one of these molecules; those with the rarer blue-yellow color blindness, with a different molecule.

In octopuses, however, light-sensitive cells in the retina contain only a single pigment, which has peak absorption of light in the blue wavelengths. With only a single pigment in the retina of the octopus, they cannot perceive color by the relative responses of receptors with varying sensitivity, as do many animals with color perception. Dogs have two types of photopigments and have less color sensitivity than primates. Parrots are tetrachromes (four types of photopigments) and their plumage may appear even more resplendent to them than it does to us. Octopuses, with monochromatic vision, do not differentiate color using this mechanism.

Scientists have speculated that the cephalopods may make use of completely different systems, using either filters to convey color information to the brain or using small differences in how in-focus an image is on the retina of the eye (chromatic aberration).

The skin of octopuses contains the same photoreceptive molecule found in the eyes, and thus is light sensitive, as demonstrated by the arm's ability to retreat from light. The skin also contains chromatophores of different colors, those same structures whose voluntary expansion and contraction determines the mottled, disruptive, or uniform body patterns. With chromatophores as color filters, and the pigments as skin photoreceptors, if the brain integrated these two sets of knowledge, color information in principle could be available to the animal.

In the octopus eye, alternatively, wavelengths of light from blue to red are in focus through a lens at slightly different distances. This variation by wavelength (color) is called chromatic aberration and may be familiar to photographers. Possibly the horizontal bar pupils of octopuses, and the even weirder U-shaped squid pupils and W-shaped cuttlefish pupils, increase chromatic aberration in different parts of the image, maximizing the difference between, for example, how out of focus red parts of the image are when the blue parts are sharply in focus. Octopuses use muscular movement of the lens of the eye to focus the image on the retina. Again, if the sharpest color in the image on the retina were detectable along with the muscular movements of the lens of the eye, the brain could combine this knowledge to obtain color information. However, we do not yet know whether the needed information ever reaches the brain for filter perception or chromatic aberration perception of color.

In addition, no known behavior of an octopus demands color vision, although people have tried to find such behavior. Experimenters tried to teach octopuses to discriminate based on color alone, but failed—octopuses apparently cannot be trained to do so, which itself is evidence that they do not, in fact, see color. Further, cuttlefish do not attempt camouflage against a background where shapes differ only in color and not brightness or other visual characteristics. The camouflage behavior of cuttlefish is based on the brightness, size, contrast, and shape of the surroundings, but not on color itself. The same appears to be true of octopuses, as experiments with *Octopus vulgaris* were able to train animals to discriminate based on each of these except color. All the evidence suggests that most cephalopods lack not only the pigments typically used by animals to discriminate colors but also any ability to use color information to alter behaviors.

The second striking fact about octopus vision is that, despite not seeing color, octopuses can see the polarization of light. Our own human eyes do not. Polarization vision can be useful in the sea. Sunlight starts out unpolarized. Unpolarized light oscillates at all right angles around the direction of travel. Light is polarized when all the waves vibrate in only one plane, like waves traveling on the water's surface. Flat shiny surfaces, such as glass, a flat calm sea, or silver fish scales, all polarize light to different degrees. Light waves in the glare off a lake or a road, once polarized, rise and fall in the single plane of the reflecting surface. Polarizing sunglasses absorb some horizontally polarized light, but let light on other planes through, thereby cutting glare from water and other flat surfaces.

Small particles in water scatter the sunlight as it enters, and undo any former polarization. Underwater, light reflected off polarizing surfaces oscillates in sharp contrast to the unpolarized background light. Octopus photoreceptor cells have alternating arrays containing the light-sensitive pigments. The cells are long and thin, and the arrays are oriented at right angles to the cell length and to each other in a regular receptor arrangement. The regular arrangement at right angles to each other permits the octopus eye to identify the plane of polarization in the light. With the discrimination of polarized light, the octopus gains detailed information about the environment. Polarized light reflected from predator and prey stands out from the background.

........................................

ALTHOUGH IT DID NOT EVADE its approaching predator, the scallop in the opening event can also see. A scallop has not two but dozens of beautiful eyes—up to two hundred, each about the size of the head of a pin, peering out between the tentacles of the mantle along the edges of their shells. The eyes of a scallop have an image-forming mirror and a pair of stacked retinas, features unlike those of any other eyes in the animal kingdom. The concave mirror focuses light. Scallops, some crustaceans, and the Brownsnout Spookfish (*Dolichopteryx longipes*) are the only animals known to use a concave mirror instead of a lens to focus an image in the eyes. The spookfish's mirrored eyes look in two directions at once. A conventional lens focuses light from above on a large retina. In the same mirrored eye, a smaller downward-facing lens bounces light off silvery tissue into a secondary retina located to one side.

What does the scallop do with its mirror and two stacked retinas? The precise functions of this anatomy remain challenging to optical biologists. The upper (distal) of the two retinas has in focus the upper part of the scallop's visual field and responds to the sight of a distant moving object. I have found it almost impossible as a diver to approach a scallop without it shutting or swimming away, even when I am careful not to change the light falling on the scallop itself. My shape is part of the mirror-focused image reflected onto the distal retina of the myriad eyes. If I were a possible predator, the scallop could take defensive action. However, the scent or touch of a starfish predator much more readily elicits swimming escape behavior than does my nearby movement.

Scallops also use vision to decide when to feed, opening their mantles to filter feed when they view particles of the right size floating by in steady currents that are not too fast. Other aspects of scallop eyes are familiar, such as the ability of their pupils to constrict in bright light and fully dilate in the dark, a capability shared with octopuses and humans.

The lower (proximal) retina also has a role in dim light, when the pupil dilates. The image in focus here is from the lower half of the scallop's visual field, perhaps the view of the substrate from the perspective of a swimming scallop. The lower retina may also help a swimming scallop detect preferred habitat, to control movement into brighter or darker areas, or to decide when to stop swimming.

Scallops are able to direct their swimming at least somewhat. The swimming species of scallops have more capable vision than nonswimming scallop species. Optic lobes are present in the nervous system of scallops but not those of other bivalves. The optic lobes are specialized visual centers of the scallop nervous system, and they process the focused images on the two retinas. The size of the optic lobes in a scallop species increases with the number of eyes characteristic of that species. The left optic lobe processes images from the eyes of the upper valve; the right lobe, images from eyes of the lower valve. The anatomical positions of the eyes around the scallop mantle are preserved in the positions within the optic lobe where their nerve signals arrive.

Scallops orient themselves on the seafloor in response to visual cues or water flow, for example to pointing downstream when beginning to swim. When swimming, scallops sometimes achieve a direction of motion independent of current flow, to either approach desirable habitat or move away from a threat. The distance at which they can do this, based on sight, may be limited to a foot or less, even when visibility in the water is greater. Scallop species vary in their shell architecture and swimming ability, and the swimming of scallops can be surprisingly fleet and linear. When above their chosen end point, they stop clapping their valves, and the scallops slew to the bottom at rest. Whether or not the scallops still see their pursuer at this point, the pursuers can still see them—if the predator is fleet enough, such as an octopus or a fish.

Octopuses have two eyes like people do, but while human eyes face forward, allowing binocular vision, octopus eyes look to either side. They have little to no binocular overlap of the visual range of the two eyes in front or behind, and as such octopuses lack the binocular depth range vision that we enjoy.

The view received by each eye informs the body pattern display on the ipsilateral side of the body: that is the left visual field informs body patterns on the left, and the right visual field informs the patterns on the right. It is easy to evoke effects of this in an octopus with a brightly lit white visual field on one side and a dimly lit black visual field opposite. With this light-and-dark visual input, the octopus will turn pale on the bright side and dark, almost black, on the dark side. An abrupt border neatly divides the two halves, running from between the octopus's two front arms up between the eyes and along the length of the mantle.

Octopuses see widely around them, such that it is difficult to tell where their attention lies within the visual field, as they do not have to turn their eyes directly toward an object of attention. Still, the eyes can move a bit forward and back. Beyond these general facts, however, there is little more known about the field of view of octopuses.

How does an octopus know how far away an object is, given their near lack of binocular overlap of the visual field of the eyes? One possibility is that the octopus may get a sense of distance as an approaching object becomes more sharply polarized as it gets nearer. This is much like humans, for example, getting a better sense of the distance of a ship as it approaches through the fog. Another possibility is head bobbing—the vertical raising and lowering of the octopus's head. As far as I can find, little to no data have ever been examined about when octopuses do this and why. Still, the expectation among those who work with octopuses is that this is a ranging behavior. By moving the eyes and the top of the head, the octopus can generate parallax within a single eye from the two positions, up and down. This may permit gauging three-dimensional information like distance, much as we do using the parallax arising from the distance between our two eyes, left and right.

Let's return to the example that opened this chapter and add some details. When the scallop first clapped up into the water, the gloomy octopus oriented his right eye toward the swimming prey. As it swam up, and when it was already wildly beyond the reach of the octopus, the octopus nonetheless reached out with his first right arm as though to grab the scallop. When this failed, the octopus bobbed his head over a span of about two inches, up and down, once. Twice. Then came a hesitant second reach by the first right arm, quickly aborted. The scallop was out of reach. Only then, did the octopus begin to move out of his den to approach the scallop as it landed on the sea floor.

Perhaps the first reach failed because the octopus lacked the depth perception necessary to gauge how far away the swimming scallop was. He learned this from parallax information requiring only two head bobs; the abortive second reach might have begun in anticipation of the scallop landing nearer instead of out of reach. Finally, distance estimated, the octopus made a quick foray over to the resting scallop, surrounding it with two arms.

In savannah-like habitats, the advantages of planning based on vision are at their peak. In these open terrestrial environments, moving animals are alternately revealed in the open and hidden by cover. The long visual range in the clear air permits both predator and prey to improve their chances when considering possible scenarios. In the long-grass plains of Serengeti National Park in Tanzania, gazelles raise their heads to keep an eye out for predators. The cheetah are careful to target the less vigilant gazelle individuals as they stalk closer to a herd. The much shorter warthogs, however, spend little time looking for predators, because they are not tall enough to see over the grass. The visual environment from their low perspective does not extend far. The advantages of plans based on the view may be an important selective pressure in the cognitive evolution of animals.

Visibility in underwater habitats where octopuses dwell is limited, however, by the spatially complex structures of reefs or kelp forests as well as by particles clouding coastal and temperate waters. This limited visibility may restrict the value of planning, because distant events that might take time to arrive are not visible underwater where only near objects are in sight.

I have been sixty feet down to the accumulated scallop shells on that languid midday on the seabed. I went to retrieve the camera on which colleagues and I recorded video of the gloomy octopus catching the scallop for lunch. In these same videos, I can see the octopus at times alerting to an approaching animal well before I can resolve anything at all in the video. Both the camera and I lack the ability to see polarizing information in the light, but with this ability, the octopus sees farther into the underwater murk than we do. Octopuses must react quickly to reach toward possible futures—whether to flee, where to land, and when to intercept others approaching. They do this with exquisite sensitivity to the context.

# 13

..............

# Reaching Octopuses

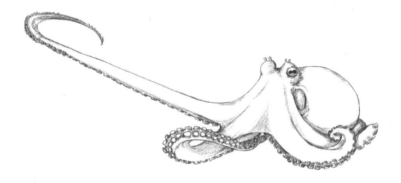

Cordova, Alaska

The octopus was kept in a shallow broad tank that hummed and gurgled in the back of the marine biology classroom. She was a two-year-old female named Amy (a diminutive of Amiguk). My brother, Ed, had brought his family to Cordova to visit. My niece, Alex, was three, and we took her to see Amy in her aquarium at the Cordova High School one day after school was out. The octopus fascinated Alex, who watched while I played with one of Amy's front arms in the cold water. Ed petted the smooth wet skin of the octopus and endured her icy grasp as she gently explored our fingers and hands with hundreds of squat inquisitive suckers.

"Want to pet the octopus?" Ed asked the little girl in his arms.

Alex emphatically shook her head no, smiling at the absurdity of the notion. However, a few encouraging words from her dad and Uncle Dave put a new face on the idea: maybe we were serious—maybe we would *make* her touch that octopus. The smile faded and was replaced with a shrieked "No!" and Alex buried her head in Ed's shoulder. We let it go at that, and Alex's good mood soon returned. She never did get interested in touching that cold tan coil of suckered flesh in the tank.

Alex's reaction to Amy was normal. Upon first meeting cephalopods, almost anyone is hesitant to touch. There is a twinge of revulsive horror at the sinuous arms, which can evoke in the human mind an ancient circuit—wiggling like worms, or worse, slithering like snakes, and slimy. There is no recognizable face. In my first experiences with octopuses, I wondered: Just how fast could those serpentine arms of ropey muscle whip out, coil tightly around me, and drag me into the empty dark depths? A good question, and Alex and I are not alone in this first reaction to octopuses. In the words of Victor Hugo, "The Unknown has power over these strange visions, and out of them composes monsters."

### August 1903, off Victoria, BC

Captain S. F. Scott was out yachting with friends for pleasure, but he was still about to have a bad evening. He had taken a rowboat out alone, a mile from his friends on the yacht. He was surrounded by a school of black-fish (that is, in modern parlance, a pod of orcas). One of these struck the rowboat hard enough that Scott was flung into the water. Even so, the orcas around him were not going to be his biggest concern.

On landing in the water, Scott was amused by the mishap. He swam back to his rowboat, and grabbed the upturned keel. At that moment, he was seized below the knees and jerked downward with such force that he flipped the rowboat over on top of his head.

"An octopus!" Scott immediately realized, as these large animals were well known in the area. Kicking hard, he momentarily freed himself and renewed his grip on the now upright boat. The octopus again grabbed his legs and pulled downward, as Scott desperately clung to his only support. He described the pain from being pulled as "excruciating"; but after long moments, the grip lessened slightly. Kicking with heavy boots and twirling to finish breaking the octopus's hold, Scott freed himself but he was badly injured.

Sometime later, his yachting friends noticed the motionless rowboat and hurried to find him half dead from injuries. His skin had been largely stripped off from feet to knees and his upper legs soon bruised black. His recovery took him the next seven months.

Here was yet another story, an addition to the tales of the hundred-pound octopus that a fisherman could not pry loose from a boat it had grabbed and that had to be cut off, of those octopuses that caught and ate birds and dogs, of the octopus that held a man underwater for two hours. I heard stories from Sugpiaq, Eyak, and Haida Alaska Natives about octopuses with touch so delicate that they could pick blueberries one by one, and a grasp so strong they could seize a man or a woman and carry them to their octopus homes beneath the sea, or reach onto shore and pull down houses. A fisherman told me about a large octopus on deck that stole a knife from its sheath on the man's belt, and stabbed him with it before escaping. Something about the way an octopus looks and moves inspires a trepidation about just what an octopus can do. How strong are they? What are their normal movements? How do they put their unusual form to use in their day-to-day lives?

Captain Scott's story raises these same questions. What does the octopus perceive as humans and orcas disturb the waters above its den? The capabilities of octopuses, their eyes, arms, and suckers, are on full display in this vignette as the octopus overwhelms her visitor in a moment.

Captain Scott had about eighty-six billion neurons in his brain, more than 95 percent of the total neurons everywhere in the body. This proportion is typical of vertebrates. Octopuses are different: neurons in the octopus's brain do not comprise the largest part of its nervous system. Two-thirds of the neurons available to an octopus are distributed throughout the body.

What do octopuses do with so many neurons outside the brain? Consider the simple motion of the arm that, in reaching up from beneath broad kelp blades, wrapped around Captain Scott's legs. The unfurling of the arm in a reach is governed within the arm itself without necessary reference to vision, despite that the largest parts of the octopus brain are the optic lobes, which process vision (among other things). The optic lobes contain sixty-five million neurons each, or about 25 percent of all neurons in an octopus body. All the rest of the brain contains only another 8 percent of the neurons, for a brain total of about one-third. Although an octopus has slightly more neurons in total than a cockatiel, about as many as a starling or a rabbit, and twice as many available as a rat, there is a big difference between the vertebrates (including humans) and the octopuses in the proportion of processing power distributed throughout the body.

The unfurling of the octopus arm has little necessary reference to the brain at all. Many neurons outside the brain lie in the arms, along central axial nerve cords running the length of each arm and in the connected chain of ganglia associated with each sucker and each pair of suckers. The result of all this processing power in the arms is that octopus arms can achieve much without requiring the brain. The arms unfurl in graceful reaching motions or sweeps that the brain may cue but that are orchestrated entirely within the arm.

The number of coordinates required to completely specify motion are the degrees of freedom of a system. The human arm has fixed degrees of freedom—the shoulder and wrist rotate vertically and horizontally, the elbow bends and rotates. However, the octopus arm bends, twists, stretches, and contracts, and can do so at any point along its length—nearly unlimited degrees of freedom. To control the motion of this limb, one that can have, if not really *any* shape, still nearly infinite shapes, the biology of the octopus must somehow simplify this infinity of specifications.

The distribution of neurons throughout the body is evolution's answer to the octopus's neurological challenge arising from the lack of a hard skeleton. By contrast, familiar animals, including humans, other mammals and all the vertebrates, and the arthropods (insects, crustaceans, spiders) have joints organized on the principle of levers, and contain opposable paired muscles to move them.

But there is no hard skeleton in the octopus's arm. Instead, the arm acts as a muscular hydrostat, in which both support and power for movement are provided by contraction of opposing muscle groups. Transverse muscle fibers, when contracted, narrow and elongate the arm. Longitudinal muscle fibers can shorten the arm, opposing the transverse muscles. Oblique muscles that spiral around the arm in either direction allow for twisting, as well.

Much of the control is done in the arm, not the brain, via a few simplifications: first, the arm behaves like a snapped whip, so that bends are propagated along its length, rather than needing to be worked out for each position they must successively occupy; second, the arm forms pseudo-joints, reducing complexity as though there were a rigid skeleton.

To extend an arm like a whip, the octopus twists the arm in a bend, for example about midway down the arm. Then the arm curves between the twist and the body, drawing the bend toward the body, so that a single curl forms. This curl is then propagated down the arm just as a whip is snapped, although at slower speed. When visually guided, the wave travels along a line between the octopus's eye and the object it is reaching toward. The arm extends to the object as the wave reaches the arm tip. When a second arm also reaches, as the gloomy octopus did catching the scallop for its lunch, as related earlier, the second arm uses the identical motion as the first arm.

To pull an object toward itself, the octopus uses similar dynamics to bend the arm at a temporary pseudo-wrist and pseudo-elbow. The first (most distant) segment grasps the object and a bend forms. Waves of muscle activation begin at this spot and at the base of the arm, and move toward each other. The pseudo-elbow forms where these waves collide. This temporarily divides the rest of the arm into two sections of equal length, and thereby approximates the proportions of an arm of a primate. Lacking joints, the octopus makes them where they are needed, and discards them when done. The octopus can now bend its (pseudo-) elbow and twist its (pseudo-) wrist to do things like tuck a tasty scallop to its mouth, or pull Captain Scott by the legs below the surface.

Another way that the octopus accomplishes so much outside the brain is by using the environment to aid its search. Octopus arms are directed by the contours the octopus encounters while searching, allowing the arm to explore surfaces, again with minimal central control. Octopuses are more successful when searching by touch along a surface's contours than when blindly sweeping an arm through an open space. When a sucker on a surface encounters something interesting, one sucker recruits adjacent suckers to further grasp and investigate the object. This process repeats with the next sucker, and so on, until the arm has a firm grip on the subject of interest, be it a bit of food, a curious object, or an unfortunate swimmer's legs.

Is the octopus thinking with her arms? The brain does not need to trouble with what the arms can do themselves. The suckers hold by default, and pulling is standard practice to discover what has give and what does not. Letting go depends more on central control than holding and pulling.

Victor Hugo likens an octopus to an umbrella with no handle, and indeed, an octopus moves like an animated umbrella. She dances in the water, her flesh flowing like the finest shimmering gown that embraces the belle of the ball, arms turned in galactic spirals, each trailing a billowed scarf of her web.

Octopuses in motion are intrinsically graceful. It is also challenging to see how they move. The body itself seems not be involved in the motion, which is accomplished by the arms alone (except for jetting). Yet with eight arms all moving at once, and each arm studded with hundreds of suckers also in motion, no detail leaps out as the mechanical force behind movement. The resistance of the water limits the speed at which an octopus can move, so that even the most mundane action, every step, appears effortless and imbued with dignity. An octopus glides along like one impelled by gravity or wind alone, but this is illusion. It is the way that a waltz deceives the eye, drawing it away from the feet and toward the syncopated rhythm of swirling gown and the sway and pulse of the music. It is the impossible display of lithe power and flexibility of Olympic gymnasts. It is the illusion of Michael Jackson, repopularizing

Apollo Theater dancer Bill Bailey's "moonwalk," stepping forward but pushing backward.

The octopus's eight arms are arrayed in a ring, with radial symmetry rather than the bilateral symmetry evidenced by the eyes. We refer to the octopus's arms in a bilateral arrangement as right and left, and in numbered pairs one through four, beginning with the pair just below the eyes and ending beneath the mantle.

The octopus's lack of skeletal rigidity and the complex musculature and neural anatomy, along with elements of both bilateral and radial symmetry, seem to offer a limitless potential of motion and pose. Indeed, octopuses crawl in any direction relative to their own body symmetry, and often prefer to crawl at a 45-degree angle rather than straight ahead. To achieve its directional choice, one or a few arms opposite the direction of movement push an octopus along. Each involved arm pushes by elongation, and each pushes simultaneously with equal force. Each arm tip stays within its own octant, as octopuses seldom cross their arms when crawling. The arms each push from a fixed angle relative to the body. The direction of the achieved movement depends only on which arms are involved in pushing.

With eight arms to choose from there is no discernible rhythm to an octopus's gait, making it quite different from the distinct metronomes of, for example, a horse walk, canter, and gallop. Bilaterally symmetrical animals evolved neurological central pattern generators that drive efficient and constantly repeating gaits. Octopuses lack this repeating pattern and instead use the radially symmetrical and fixed distribution of eight arms around the body to direct movement.

The complex and mysterious flow of an octopus traversing the ocean floor is a series of motions of the arms and the sucker stalks bending and straightening, extending and contracting in length. Each sucker may grasp, relinquish, or hold any surface it contacts. Finally, octopuses flare and retract the thin web of skin between the arms. The flow of water against all these surfaces creates additional motions of billowing and flapping like the canvas and lines of a clipper running before the trade winds.

Objects can be carried beneath the web. We might expect, with eight arms, that an octopus would carry things in some arms, and use the others for walking or crawling. However, this does not appear to be the case. Carrying objects in extended arms would off-balance an octopus, and instead, objects are carried close to the center of mass, which is near the center of the arm crown.

When another object is to be controlled, an octopus will try to envelop it in the web and arms. This is visible, for example, when two male octopuses fight. The aggressor may grab with one arm, pulling his opponent under the arms and web. Each rears back, expanding the web and arms widely, reaching to get over the arm crown of the other. If successful, the opponent is enveloped. To flee, an octopus must escape the grip of suckers, and adopt a streamlined pose to jet rapidly away. Sometimes this is done from within the grip of another, but sometimes the octopus, successful in enveloping the other, will push itself into a streamlined posture, releasing the opponent from his grasp at the moment he jets away.

When returning to the den after successful foraging, an octopus will carry the prey just forward of center, beneath the web. The distal parts of the first arm pairs may still be exploring or crawling; and arm pairs two, three, and four can be involved in crawling, walking, or even jetting forward. With suckers along the entire arm length, octopuses can hold something close to the body and still put the outer reaches of the same arm to other uses.

Octopuses also sometimes gather material they would like to carry. To do so, the material is gathered by the arms and held in a bolus that's wrapped in the web. The arms and extended web curve over the mass of gathered items or material, coming together underneath it in a pinched-off closure below which the arms again bend outward to the substrate. The appearance is of a walking balloon with tiny little suckered legs below, and eyes and a mantle above. I suppose an octopus might carry food this way after very successful foraging, but more often I have seen this behavior used after meals to carry remains of the prey slightly away from the den before discarding. The distal third of the arms below the inflated ball of web are curled into pseudo-joints and the octopus walks forward on these to its destination. The movement of such a large dirigible of head, mantle, and inflated web, by such abbreviated limbs below, looks improbable, and at that moment, an octopus so occupied can look much more comical than threatening.

# 14

...............

# Sensational Octopuses

Octopuses like to hold hands, and once one gets past any nervousness that the octopus means harm, it's endearing.

For an octopus, holding hands with a person is an active endeavor, because of the many suckers involved, each of which is holding on some of the time, but also busy exploring. Octopuses seem enamored of the feel of skin, and unless there is something noxious on a hand, the many suckers seldom let go all at once. Instead, one by one, a fingertip or knuckle passes to a neighboring sucker, and the octopus arm itself wetly ascends the hand to the wrist. Crevices are interesting—what is under the cuff? Soon the leading part of the arm has crawled into your sleeve.

The explorative curiosity, like other abilities, seems embodied in the suckers and arm. I recall earlier in my studies when I had gone subsistence foraging out of Chenega. I watched Mike Eleshansky clean a just-captured octopus for his supper. We were standing on the float that made up the small boat dock. I had seen other people, when butchering an octopus to cook, slit open the mantle, remove the internal organs, cut out the beak and associated muscles, and keep the meat of the head, mantle, and arms to eat. Wild foods were routine for Mike, and he went at it more simply. He sliced each arm off the octopus at its base and tossed them one by one into a bucket. The head and mantle he discarded in the water. Just as he was finished, another man came along the float and chatted with Mike.

"How'd it go?"

"Got this one. Twelve pounds."

They talked for a while—about fishing, the weather, family. Casual talk. As they chatted, I was fascinated to see the arms of the octopus, despite their separation from the head, feeling their way out of the bucket. They crawled tip first up the white plastic side. The first one made it several inches over the rim of the bucket before Mike glanced down and casually peeled it back sucker by sucker and dropped it to the bottom. The tip of the arm commenced the journey again, marching up sucker by sucker. Periodically throughout the next twenty minutes of conversation, Mike would reach down, peel loose the most advanced escaping arms, and toss them back, from where they would again begin anew.

HOW ARE WE TO UNDERSTAND such absent-minded activity? Until the severed base of an arm came into view, the scene in the bucket was exactly that of an intact octopus bent on escaping captivity. The presence or absence of the animal's brain was unimportant. The arms were crawling out of the bucket, determined to return to the sea.

In Tlingit tradition, there was a time before there were any animals in the sea. The apparent purposefulness of octopus arms that I saw in the bucket also features in this story, in which Raven carves canes into the form of two octopus arms. This is the story of how Raven filled the sea with life. As Raven reached his carved octopus arms toward a distant floating object—an everlasting house filled with life—I imagine the arms must have grown longer and longer, reaching on their own, seeking their target. At that moment it appeared the arms themselves sought to bring the everlasting house to shore, although really this was Raven's purpose, and the carved arms were his tools. In much the same way, on the dock in Chenega, octopus arms crawled up out of the bucket on their own, revealing embedded intent and purpose deep within their anatomy.

The reaching and fetching of octopus arms is coordinated within the arms themselves, and is done even absent the brain. So, too, are the actions of the suckers coordinated locally. Each of the hundreds of suckers is a muscular hydrostat; like the arms themselves, each is capable of movement in any direction. Each sucker contains pressure receptors that signal when something touches the sucker. Suckers connect to each other through ganglia in the axial nerve, and these coordinate signals from one sucker to its neighbor, that turn and reach for the surface touched by the first sucker. The brain is not involved.

These arm-local actions can be purposeful. For a short time, I was collecting biopsies from octopuses for an early study on the genetics of the giant Pacific octopus. Each biopsy was taken as a small amputated arm tip from a live octopus. While I disliked the necessity to take tissue for this study, partial arm amputation is common among octopuses in the wild and is less severe an injury than amputation in vertebrates (see Chapter 11). I preserved each biopsy by dropping the arm tip into a small vial with pure ethanol. Volatile and harsh-smelling fumes arose as soon as the vial was opened. The moment a biopsy encountered the fumes rising off the ethanol, the arm tip would twist and turn away. It curled itself outside the opening of the vial. The suckers reached for any alcohol-free surface.

Before the genetics study finished, we developed a better technique. We collected mucus to study octopus genetics (and eventually hormones), by gently rubbing octopuses on the head or mantle with a lab-grade Q-tip. This reduced the harm inflicted on octopuses, yet still allowed us to learn more about them. I was relieved that I did not have to take any more biopsies. Not only did I dislike injuring (however mildly) each octopus to obtain a biopsy, but I felt bad for the arm tips struggling to escape preservation in the alcohol.

These escape actions of the arm reveal that judgments are made locally about whether an odor is enticing or aversive, based on information from receptors somewhere in the suckers or skin of the arms, and in this case at least, without any reference to the brain. When something noxious stimulates sensory receptors, the suckers will turn away from it, or if it is a contacted object, will drop it immediately. Occasionally a stimulus that is unattractive but not noxious will pass in a conveyor belt of suckers out the length of an outstretched arm to be dropped away from the animal. More often, when the receptors are stimulated by a neutral or attractive object (one that tastes like crab for example), the object is passed from one sucker to the next up toward the mouth.

Arms also in the same way avoid pinches and contact with fresh water or with vinegar, but do not avoid seawater or gentle touch. The suckers bear exquisite sense organs that respond to light, scent, taste, touch, and pain, as well as to location, providing the information necessary for these judgments. Sensory information from suckers evokes behavior of the arm, via the adept neurology at work within the suckers and axial nerve cords. The senses of each sucker determine within the arm decisions about recruiting neighboring suckers, passing an object out to the end of the arm or toward the mouth, or avoiding objects entirely.

Octopuses have a sense of smell that, judging by the reaction of the arms to the alcohol fumes, involves local chemoreceptors. These may be specialized to detect soluble molecules in the water. As indicated by the ability of an arm tip lowered toward the vial to move *away* from the fumes, the arms can gather information about the relative location of the odor source.

In the dark, without vision, the octopus spreads out. Octopuses expand and contract the reach of their arms to investigate a scent of interest as they move toward the source of an odor plume. They redirect their path as the extended arms detect concentrations of the scent that drop off as they reach the edges of the odor plume.

The densely packed sensory cells along the sucker rims include the cells mediating touch and touch-taste. The sense of touch depends on receptors that sense pressure and vibration. Touch-taste depends on receptors that detect molecules that are on the surfaces touched by the suckers. The touch-taste receptor cells are unique to cephalopods and not found among other animals, although their function is somewhat like our taste buds.

Octopus arms delve into crevices, each exploring arm creeping as far as possible blindly farther and farther into the space, searching. Possibly the arm will find an encrusted bivalve glued to the rock, or a crab to capture, but possibly instead it will find nettlesome dangers like stinging anemones, sharp urchins, or abrasive sponges. Touch-tasting suckers at the point of contact can determine whether to adhere or release, to advance or retreat—an efficient, fast way to avoid dangers or catch prey that may flee on touch.

AT THE BOTTOM OF MIKE'S BUCKET on the float in Chenega harbor, each sucker passed its piece of the white plastic floor to its neighbor, who in turn forwarded it on toward the mouth. Only the arms had been detached and the mouth was not there. Instead of moving the bucket, the coordinated action of all the suckers moved the arm itself, tip in the lead, across the bucket floor and up the sides toward the sea. The zombie determination I saw was not my own imagination at work, nor was it experienced by anything in the octopus arms themselves. It was the evolutionary embodiment of the wisdom of the octopus.

Later that day, Mike boiled the octopus arms in a big pot of salted water on the stove. After they boiled, he placed them in a bowl to cool, and when they were cool, he sat in a chair in his small dining room at the table. I sat opposite him to watch. He stripped the dark reddish-brown skin off each arm. The suckers came off cleanly with the skin, leaving behind a tapered curving cone of startlingly white meat, an edible tusk.

"Know why we strip the buttons off?" he asked, indicating the pile of discarded skins buttoned with suckers.

"No. Why?"

"Else when you eat 'em, they crawl back up again!" Laughing, he sliced each peeled arm into disks like a banana. We ate the clean ivory flesh with a dash of salt. It tasted of the sea.

DOES THE OCTOPUS ITSELF have to contend with the looming autonomy of its own arms? The potential for that autonomy is clear from the actions of the detached arms in the bucket. Do the arms ever engage in behavior the octopus seems not to intend? How could we know?

Earlier, I told the story of a gloomy octopus off Australia. Sixty feet down on the shell bed, it twice captured then dropped its scallop lunch. It was as though the visual system of the octopus recognized the scallop as a desirable meal—so it picked up the scallop and took it back to the den to eat. Once out of sight under the web, the tactile-chemical system found the scallop unpleasant or uninteresting—and the octopus let it go. Not once, but twice. How could this be? Did the gloomy octopus know what its arms and suckers were doing?

These foraging incidents seem mysterious, but they might be more understandable given the way the octopus nerves and brain are built to work together, and with a little understanding of one more tactic the scallop has to fool an octopus.

Many scallops have a shell encrusted with sponges that often completely cover at least one side of the bivalve. The weight and drag of an encrusting sponge make it harder for a scallop to swim. So why do the sponge and the scallop associate in this way?

The sponge benefits when the scallop swims; this motion can shed predators such as sponge-rasping dorid sea slugs. The scallop benefits, too, by hosting the sponge. This has been demonstrated in captive studies where scallops were attacked by sea stars (such as the sunflower star, *Pycnopodia helianthoides*). The star's tube feet stick by a chemical glue, which does not adhere as well to the surface of the sponge as it does to the hard calcium carbonate surface of the scallop itself. In this way, the swimming response of the scallop can dislodge an attacking sea star. In captive studies, scallops encrusted with barnacles rather than sponges did not escape sea stars more easily than did those without any encrusting animals at all. However, it is uncertain whether sea stars are major predators of scallops in the wild, as scallops smell and see them coming. Scallops in the wild can swim away before any contact occurs with an approaching star.

Many species of octopus, including giant Pacific octopuses, red octopuses, and the gloomy octopuses south of Sydney eat scallops. Octopuses have a long evolutionary history preying on scallops. To an octopus, scallops look like food, and their swimming behavior draws the eye. When given both sponge-encrusted and bare-shelled scallops, however, octopuses are two to five times as likely to feed on the bare-shelled prey. Octopuses, then, are a source of natural selection allowing a sponge-encrusted scallop better odds than a bare-shelled scallop at surviving an octopus encounter.

The indecisive gloomy octopus profiled earlier, seeing a swimming scallop, captures it for a meal. Once captured, the prey left the visual system and passed under the web, to the chemo-tactile system. Here it was identified by its surface, and this scallop was heavily covered with sponge, which octopuses don't eat. The well-adapted suckers therefore released the scallop, which was then free once again to be seen swimming past the octopus, who can recognize an easy meal when it sees one!

Given that octopuses often forage with their arms, for prey out of sight within crevices, it is not surprising they prioritize chemical cues over visual in choosing food. In the case of the confused gloomy octopus, the scallop moved in and out of sight, and the chemical cues deceptively conveyed a message of unsuitability. In these unusual circumstances, the cues for prey choice cycled between the view of a suitable food item and the touch of something unappetizing.

Octopuses rely on vision for some aspects of their lives; but for the rest, they spread out, touch, taste, and grasp to encounter the world. The nonvisual sensory world of the octopus is important, and some animals that octopuses seek, such as the scallops, have evolved to be nearly unrecognizable in this grasping worldview. Still mysterious are the important ways that the octopus's nervous system spreads throughout the arms, suckers, and skin, and how these adaptive ways of functioning interact with the brain. Such questions can and are being studied, but there is still room for researchers to disagree on these matters, and clear answers take time to emerge from their work.

# Octopus Cognition

# 15

## Constant Octopuses

**Anchorage, Alaska**

"Dena has become feisty."

One of the student aquarists stepped into my office to report a change. Dena was usually very shy, almost reluctant to interact with anyone. Now, Dena interfered with aquarium cleaning and squirted the aquarists. Worried about the welfare of my charges, I wondered what was going on. Dena, of course, was an octopus.

I began my octopus studies in the coastal fishing town of Cordova, Alaska. I continued them, however, from my appointment as professor of marine biology at a private university in Anchorage, Alaska. The waters of Prince William Sound were not as accessible from Anchorage, on the traditional land of the Dena'ina people, but I was nonetheless able to work with many capable and interested students on the curiosities of octopus behavior.

My students and I had designed and built a thousand-gallon capacity aquarium facility on campus. The main octopus tank was four by five feet on a side, and provided space for one giant Pacific octopus, as long as the individual wasn't too large. A neighboring tank of four by one and a half feet was limited to our smallest octopuses. I trained interested students as aquarists, and they learned a lot of science by taking care of the octopuses and other animals in the aquariums. We mixed artificial seawater in a fifty-gallon tank, and fed the octopuses live clams from a local seafood store.

Dena's change in behavior started right after we released Calamity—an octopus in the aquarium adjacent to Dena's—back into the ocean. Eighteen months earlier, Calamity earned her name when she startled herself right up into my hands as I first investigated her den on the rock-strewn beaches of Prince William Sound. She'd grown rapidly in our aquariums, and when she got too large for our space we released her back into the wild.

In our care, Calamity continued to live up to her name, stealing aquarist tools whenever possible, squirting water across the room when her tank was open, and pulling anything she got a hold of (including nervous aquarists) as far into her tank as it or they would go. Aquarists were both amused and exasperated with Calamity—she was fun to work with but would not let anyone finish their job in a hurry.

Dena, on the other hand, was under one of the last small rocks we turned over the previous season. She came into her aquarium only one-tenth the size of Calamity. They grew in size in those adjacent tanks, shared the same water, and had a partial view of each other's space, but they had never touched. At the time we released Calamity, she was still half again as large as was the very shy and retiring Dena.

A new octopus arrived and we put her in Calamity's recently vacated space. Clade was quite small, only a quarter of Calamity's size and less than half the size of Dena. Now, with Calamity gone and Clade in her place, Dena underwent a personality change, from shy to bold, from avoiding interaction with the aquarists to an insistence on interacting at every chance.

The change was abrupt. The only relevant event that might have sparked the shift was the arrival of a new neighbor. A neighbor she could not interact with, but whom she could see. Dena now could see that she, and not Calamity, was the biggest octopus in the neighborhood. And I wondered—had she been shy to stay out of big Calamity's way? Was she now feeling her own size and exerting her own will more as she realized she was the biggest octopus around? I wondered whether this was a realistic possibility.

At that time, we knew that octopuses were solitary and asocial. But . . . did Dena have an image of herself as a small octopus, which suddenly shifted when she saw that, compared to Clade, she was a big octopus? What sort of judgments do octopuses have to make and what sort of awareness do they have?

ANIMALS ARE UNIQUELY ACTIVE multicellular agents in a world vibrant with life. Sponges pump water through their stationary bodies. Corals pulse and clench; anemones walk and swim. Worms, crabs, and lobsters, sea slugs, and squid variously burrow, creep, crawl, scuttle, walk, swim, and jet through their lives, bustling about their business. The lifestyle of every species is unique in some ways. Every animal, nonetheless—indeed every living organism—has some business to go about. That business has certain universal aspects. Every animal must eat and discard waste, must shelter from predators and other dangers, and will dedicate its time and energies to the next generation.

Nearly all animals accomplish this through actions of muscles coordinated by nervous systems. Animal senses feed information about the outside world into the nervous system, allowing muscular action in the world.

Mobile animals encounter novelty. Big Calamity's departure and the arrival of little Clade marked a significant change in Dena's visual world. In the wild, this change might offer opportunity. Mindless action would suffice if the appropriate responses to novelty were always the same, but Dena's feistiness suggested she was exploring her new perceptions. The demands of novelty—the novelty of a new neighbor or a new neighborhood—mean alternatives; if there are any, they should be considered.

Considering alternatives requires some understanding of the world. Sensing size is one part of this, and I already knew that octopuses are good at sizing up their food. It is not obvious that octopuses are able to do so—they lack the binocular vision, which primates like us use to judge the size and distance of objects. However, octopuses can do this also.

In an experiment, octopuses distinguished the sizes of target squares despite varying distances from animal to target. To do this, the octopus must factor in its own subjective viewpoint—it needs to know how far away the object is from itself. Is the object near and small? Or farther away but larger? Taking into account your own viewpoint as the observer allows *visual constancy*. Animals recognize constant features of an object despite the object's varying appearance to sensory systems that arise from the animal's particular perspective.

Such constancy is also an expression of *perceptual contingency*—perceptions are contingent on the animal's own movements and location relative to the object being sensed. To recognize object constancy, the perception of the organism must account for the fact that its sensory access to the object depends on—is contingent on—the animal's own location and actions.

Sensory worlds, then, depend in part on the motion of the animal, the sensor. When the octopus eyeing the swimming scallop raises and lowers its eyes in a head bob, the image on the retina shifts, not so much with the scallop's movement (although that is also true) but *contingent* on the bob of the octopus's eyes. Animals do not normally confuse their own movements with those in the environment. Their own movements are accounted for to build the sensory world they explore.

Animals use their own movements in such sensory judgments. Yet this sensory world develops without much attention to the contingent aspects of it. Consider an experiment that offered a group of people a new sense and a new dependency of the sense with their own movement. Participants wore a belt-and-compass device that vibrated on their waist at the particular spot that was on the magnetic north side of their body. As the wearer turned, the vibrating spot on the belt moved along the circumference to remain always on the north side. Wearing this belt over a period of weeks, the participants walked, biked, and hiked outdoors. Participants experienced the new contingency of the vibration while moving and active. This deeply influenced how they understood direction. Range estimates and accuracy improved as did their sense of space.

Interestingly, however, they did not reconceptualize a cognitive map of their surroundings, realizing past errors and correcting them, for example. Instead, participants found that their perception of space improved, stating that space was "wider," or extended beyond the range of sight. Previously, this had been a cognitive construction. Now it was felt. The perception of vibration faded, much as do perceptions of most clothing after a few moments, and instead the new sense of direction felt "direct" but distinct from other senses.

Active animals filter their own movements out of their sensory world. For example, I do not confuse my typing on the keyboard with the computer keys pushing out at my fingers, and all the while, I am completely unaware of saccadic movements as my eyes follow the words I type. To selectively ignore changes in our perceptual worlds, I, and all animals, must have and use sensory information about the actions of our own bodies.

Such knowledge about their own actions scaffolds the inner lives of animals. Our understandings of the active lives and perceptual abilities of animals allow us to infer this aspect of self-knowledge, and its broad implications. Nervous systems not only gather information about the environment, they also sense the state of the body—its positions, motions, accelerations, and needs—to coordinate muscular movement through the environment.

Animals learn from their own behavior. Experience is an interaction of the internal and bodily states, with the sense organs, arising from both the behaviors of the organism *and* from the environment. No account of experience can be persuasive without all these elements. Consider again the example of the octopus and his scallop lunch, which once passed out of the visual system to the arms, was discarded not once, but twice. The advantageous scallop-sponge mutualism generated one pattern of feedback to the eyes ("Lunch!"), and a different and noncoherent pattern to the touch and taste receptors ("Not food"). With these incoherent signals from different interactions with the environment, the octopus failed to deal successfully with the specifically touch-based camouflage of this prey.

Novelty, encountered in movement, makes demands of organisms; and coping with these demands requires one more ability, that of flexible responses. Some responses may be inflexible but adaptive. Some are behaviors modified only by perceptual tracking of the current conditions; others are behaviors shaped by learning. Behaviors may be further guided by the explicit formulation of a goal. When novel circumstances create uncertainty about the best behavior to satisfy an animal's inner state, that is when the most flexible responses are needed.

Some behaviors are insensitive to feedback. That is, behaviors achieve an effect, but the outcome does not modify what the animal does. Animals with rigid morphology or simple behaviors may be without options, and an automatic response may be the only one available. Such animal behaviors are inflexible—they are specific outputs, evolved under natural selection, that reliably reach an adaptive outcome. A classic example is the egg-rolling behavior of the nesting Greylag Geese; they will rescue an egg that has accidentally rolled outside of the nest. Konrad Lorenz removed such an egg from under the bird's bill in mid-rescue, before the goose had returned the egg to the nest. The goose nonetheless completed the entire behavior, tucking the egg that was no longer there back into its nest.

Douglas Hofstadter called this inflexibility in the face of novelty sphexishness, after the digger wasp genus *Sphex* (*sphex* means "wasp" in ancient Greek). Digger wasps excavate burrows, lay eggs in them, and provision the burrow with paralyzed insects before sealing it up. When the eggs hatch, ample provisions await the newly emerged young. Before entering the burrow with provisions for the young, the mother wasp leaves the paralyzed prey at the burrow threshold, and briefly enters without it, checking inside for things that may be amiss, then returns outside and brings in the prey.

A curious experimenter moved the prey a few inches away. Mother Sphex, emerging from the burrow, retrieved the prey, then (again) left it at the mouth of the burrow for her brief inspection. When the prey was again moved, the inspection was repeated, over and over. The experimenter repeated this forty times, and Mother Sphex sphexishly inspected her burrow forty times, never wavering from routine. The routine is nearly inflexible, mindlessly diligent. The behavioral routine is the only one easily available to *Sphex* at that stage of provisioning: on approaching the den with paralyzed prey, set the load down, inspect the den, *and then* haul the prey inside. However, *Sphex* vary in how many times behaviors are repeated before they are able to break out of the pattern—some *Sphex* break out of the pattern in fewer than ten repeats, while others exhaust the patience of the experimenter before a different behavior emerges.

Other behaviors do respond to novelty. The behavior might be a guided action toward an outcome (an aim) represented internally. This behavior can be sensitive to environmental and proprioceptive feedback. That is, the animal adjusts to local contingencies. The behavior changes or ceases when the outcome is achieved. In this case, the demands of novelty must be coupled with flexibility of responses—options among which an animal must choose.

Novelty and flexibility of response are bound by cognitive ability. According to one modeling study, terrestrial organisms are more intelligent on average than marine animals. Terrestrial animals must consider more possibilities. Air is clearer than water, and in many terrestrial environments, dangers or opportunities come into view at greater distance. This allows an organism to consider a larger choice of actions. This opportunity to choose from a number of options—the opportunity to plan—favors intelligence. Novelty demands flexibility in the course of development, as well. For example, cuttlefish raised in environments where they encountered novelty remembered learned tasks better and were more capable hunters than cuttlefish raised in overly simplified environments.

To understand the inner lives of octopuses, we must remember how their bodies, nervous systems, and multiple senses interact with the environment as they learn what things look, feel, taste, and smell like, allowing them to succeed. The octopus profiled earlier that opened the giant butter clam in Port Graham not only got a meal, but learned about the relationship between a recalcitrant part of the world and her own ability to pull, to drill, and to bite through shells.

The bivalves can be a challenge to octopuses. When a new octopus arrives in our aquariums, often the animal is used to eating small crabs, as we learn from examining its midden. We feed the octopuses shrimp and small live clams, both unfamiliar foods. Often the bivalve "clams up" and does not open, and the octopus, having no experience with clams, ignores it as food. The suckers don't recognize a bit of this shell as associated with a meal.

As a lesson, I crack the shell of the first clam or two. Now, when suckers touch the broken clam, they taste the potential food, and the suckers pick it up. The octopus quickly learns that the shell, previously uninteresting, contains a meal. The octopus begins to laboriously drill open these shells, may try chipping them, and then after a few meals realizes it is strong enough to pull them open. They accept clams after that, and I can imagine the octopus salivating as soon as it tastes the shell surrounding its next meal.

Living in the novel habitat of the aquarium, the octopus combines information from its new environment across its senses and actions, from taste to pulling to the visual stimulus of the aquarist arriving for mealtime to open the tank.

In the past two years, a few papers on disconnected topics have again evoked my curiosity about the sensory world the octopuses inhabit and the judgments they make about the world and their place in it. The previously mentioned finding that octopus arm tips retreat from light, whether or not the eyes can see it, reinforces the notion that the octopus's visual and taste-touch systems are not perfectly integrated, which was further illustrated in the anecdote of the sponge-covered scallop that the octopus found appetizing to behold but uninteresting to touch.

Animal minds do not always comprise a single well-integrated self.

Dolphins and whales sleep literally with one eye open, and only half of the brain sleeps at a time. Pigeons that learn only on one side of the body do not transfer the learning to the other side—their right eye does not know what their left eye is seeing. In some ways, octopuses seem more unified than this. Sided learning by octopuses sometimes is available to the untutored side; sleep seems to be bilateral.

Still, the autonomy of the arms from the central brain of octopuses has received some attention. Octopuses use only limited proprioception information about the positions of their suckers. Octopuses, for example, can distinguish the shapes of objects they can see, but do not learn to distinguish the shape of objects they only touched. They use only deformation of the suckers over corners or curves, for example, to discriminate object shape. They cannot discriminate the weight of an object by the drag on their arms.

The anatomy of the nervous system reinforces the idea that octopus arms have some autonomy. Nerve signals are directional, and nerves running from the brain to the arms are connected with many ganglia. These nerve centers, when combined, contain more neurons than the brain itself. It seems as though the brain can send commands to the arms, but depends on the neurological organization of the arm itself to determine the action. That control has to originate locally, within the arm. Further, the arms contain no mechanisms that allow the octopus to sense arm position. In vertebrates this awareness depends on sensing joint angles, and octopus arms have no joints.

This picture of arm autonomy became quite strong with the finding that severed arms, when electrically stimulated, could perform whole-arm actions, involving coordination of many muscles, with no reference to the brain. These include reaching, attaching, and releasing suckers, and passing desirable objects mouthward.

With this picture in place, philosophers and researchers have considered to what extent the octopus brain and the octopus arms share the same perspective on the world. Perhaps the arms could learn without the brain. The neurobiology of the arm seemed capable of supporting simple learning outside the brain. Perhaps the octopus arms had their own center of awareness, and arm autonomy was considerable.

The strongly centralized vertebrate nervous system organization—one brain, one mind—stands in stark contrast to the distributed nature of the octopus's nervous system.

The growing impression of arm autonomy, possibly controlled by separate seats of awareness, however, is tempered in two ways. First, octopuses can and do exert top-down control over their arms, for example, by guiding arm movement with visual cues. This should not be surprising, but it does counter the notion that the visual system (in the eyes and optic lobes of the brain) is not in careful communication with the touch-taste systems used by the arms to control movement. The top-down control is not just visual, however, as suckers release their tenacious hold on objects more readily when in contact with the brain than when isolated. The automatic grip action of suckers is inhibited centrally, reinforcing the notion that there is a role to play for top-down control from the brain.

Second, cuttlefish can learn the same self-control that children demonstrate in a "marshmallow test": to delay gratification now for a more preferred reward later. In this test, a child is left alone with a tempting treat, such as a marshmallow, and the promise that if the child holds off eating the treat until the experimenter returns, they can have two treats, not just one. Psychologists suspect the ability to delay gratification to reach a larger goal has important positive outcomes later in life.

It should not be surprising that animals can also delay gratification, including cuttlefish. Although the experiment has yet to be done with octopuses, I expect they have the same ability. Imagine a foraging octopus that encounters an opportunity to catch a less-preferred prey. Should it spend time on this one, or defer to a later time when it might find a better meal? If the octopus expects it can catch something better before returning to its den, then it should not waste time with the lower quality item. The higher the odds of greater reward in the future, the more it should pass up the inferior opportunity in front of it. The larger crab carapaces in octopus middens and strong signs of preference among prey types suggest that octopuses do just this.

Dena, too, may have been practicing self-control as she watched life in the aquariums, learned and grew. So, too, I suspect the gloomy octopus in Australia could learn to hold on to the swimming scallop that feels and tastes like a sponge, long enough to open it and taste the sweet flesh inside.

Dena has grown larger, and this has created new opportunities to explore, to contest with the aquarists, and to discover new avenues in the world. In doing so, she is exhibiting many aspects of self-knowledge, including learning her own positions and movement, watching her arms, and quite possibly estimating her size relative to others and practicing self-control.

# 16

## Dreaming Octopus

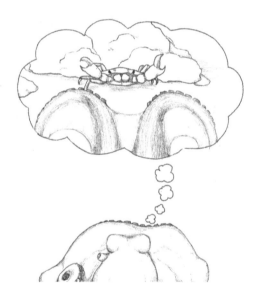

**Anchorage, Alaska**

My living room is quiet except for the hum of the aquarium pump and gurgle of circulating water. Heidi is comfortable in her home and napping, stationary and asleep on the vertical aquarium wall. She displays a mottled mixture of browns, veined with sepia and splotched with cream. Her mantle tip is dark, but lightens toward her head. Below her eyes, she is still paler. The skin over her mantle and head is scattered with small papillae. It's a relaxed camouflage body pattern, but this one is dynamic—darkness waxes and wanes over her body, as though she were moving through bands of shadow and light.

Heidi becomes dramatically darker, with ominous stripes appearing on either side for a moment. Then she pales, mantle first, then her arms. Her color is all pale. The papillae vanish. A yellow wash blooms, fades, returns, and holds. Then in a moment, her color blackens and her skin wrinkles along the veining. Pale again, then wine so dark it is black, followed by a deep camouflaging tessellated contrast pattern scattered with inky clouds. A curled arm tip twitches.

Heidi is asleep. Why do her body patterns change?

Perhaps she is dreaming. Can we know what an octopus is dreaming?

........................................

YOU SLEEP EVERY DAY. It is a period of low activity in one spot. You might roll over, but usually you are not walking around. You lie down typically; that is, you adopt a particular posture. You can wake up from sleep, although you are less responsive to disturbance than you would be when alert. If you don't get your sleep, you will be sleepy the next day and may have an urge to take a nap or go to bed early.

All animals sleep—at least, we have yet to discover cases of an animal that has no cycle of sleeplike behavior. But what *is* sleep? The actions just noted comprise the definition of behavioral sleep. They show how we recognize sleep in other animals. It is not identical to resting. Being less responsive distinguishes sleep from simply "resting your eyes" or otherwise being stationary. That you can be woken from sleep distinguishes it from other forms of unconsciousness, like being knocked out or in a coma. Sleep, unlike mere quiescence, is also regulated—you cannot skip sleep without paying the price, and you cannot skip it entirely for long. A certain amount of sleep is required for you to function.

Every species has a typical sleep posture, although these are not the same for each. Sperm whales, for example, sleep suspended vertically head up below the water surface, rising at intervals to breathe. Parrotfish sleep in shelter, ensconced in a cocoon of their own mucus. The upside-down jelly, *Cassiopea*, lives mouth up on the bottom of shallow seas. The jellies contain algae cells photosynthesizing in their tissue, from which they derive nutrition. They slowly pulse their bells in the sunshine to ventilate the algae. These jellies sleep in this upside-down position at night when their ventilation slows. Wake them, and the ventilations increase again. If kept up all night by inquisitive scientists, the upside-down jellies fall asleep more during the next day.

That octopuses sleep has long seemed the case. The demonstration of this in the scientific literature is recent though, and studies of sleep among cephalopods are rapidly increasing what we know. The first study to identify sleep in cephalopods was on cuttlefish, and reported only in abstracts beginning in 2002. Another ten years passed before a full paper appeared in 2012, showing that cuttlefish (*Sepia officinalis*) do sleep. Octopuses got attention in the same timeframe, with important studies in 2006 and 2011.

All animals likely have two phases of sleep. Consider the two stages of human sleep. First, as we fall asleep we enter a liminal state between relaxation and sleep. This lasts just a few minutes as we relax. We settle then into light sleep, and eventually deep sleep with very relaxed muscles. Body temperature drops. Our breathing and heart rate are at their lowest levels and our eyes are not moving under their lids.

At the next stage, while our big muscles remain deeply relaxed, muscles in the eyes, face, fingers, and toes twitch more. In particular, our eyes move rapidly, as though looking at something. Our heart rate increases but body temperature does not. This is rapid eye movement (REM) sleep. The twitching and eye movements make it a high-activity sleep. The early stages of falling asleep, light sleep, and deep sleep are collectively non-REM, low-activity phases. We spend up to 80 percent of our time asleep in the low activity non-REM phases, and the rest in high-activity REM sleep. We alternate between the two throughout the night.

Low- and high-activity sleep stages also occur in other animals, likely *all* other animals.

Researchers, inspired by the changing body pattern activity of Heidi when asleep, filmed medium-sized *Octopus insularis*—a species common in coastal Brazil and its oceanic islands—to learn more about octopus sleep. Often these animals were active or were alert but not active. Sometimes they were quiet but with the pupils still open—not clearly alert but not asleep, just resting.

The appearance of these octopuses was different when asleep. The pupils were closed. Skin color was uniformly pale. The researchers called this quiet sleep, a low-activity sleep phase. Short bouts of active sleep interrupted quiet sleep. The periods of active sleep lasted less than a minute each, compared to quiet sleep bouts that averaged almost seven minutes. During active sleep, the octopus displayed changing patterns of skin color and texture, as well as rapid eye movements just as people exhibit during REM sleep.

This was how Heidi was displaying when asleep. Stationary and mostly relaxed, pupils closed, but arm tips twitching and body patterns changing color and texture. She was asleep, in the active sleep phase. Animal active sleep is not identical to human REM sleep, but there are parallels and similarities.

Often when awakened from sleep, we can relate a dream experience we were having. This happens about half of the time when we awake from non-REM sleep, and about 80 percent of the time when we awaken from REM sleep. Dreaming is a common human experience, and it's more common during REM sleep than non-REM sleep.

Do animals dream? And if so, can we know anything about it? Or does all knowledge of dreaming necessarily come from people telling us their dreams, so that nothing can ever be learned about animals dreaming until, like Doctor Dolittle, we can talk to the animals? Understanding whether animals dream is important for at least two reasons. First, dreaming is a form of consciousness. If animals dream, that says something about their being conscious. Second, dreaming is imaginative. In a dream, the experiences of the dream are not related to the present environment. They are constructed—imagined—from past memories and events. An animal that dreams is one that can imagine, which is a capacity that underlies planning and creativity.

Most of what we know about human dreaming come from our own

verbal dream reports, but dream researchers use other tools, as well. Telling about your dream opens the door to knowing about it, but not *everything* that we know about dreaming comes from telling the dream.

A dream is a mental experience that occurs to the dreamer during sleep. It is subjective—only the dreamer has direct access to the dream. Dreaming is not cut-and-dried distinct from other kinds of subjective experiences such as daydreaming or mind-wandering or psychedelic experience.

Odd contradictions would be inherent if dreaming is defined only in relation to telling someone about it. If I tell no one of my dream, would that mean it didn't happen? Conversely, if I lie (even to myself) that I had a dream I did not experience, would that mean I really *did* have the dream? If I have a dream that affects my mood, but forget the dream before I can even tell it to anyone, my altered mood persists. The dream cannot depend on the telling. Rather, the reverse is true. Indeed, the experience in a dream and the memory of the experience depend on different parts of the brain. That suggests we can know something about dreams without the telling of them, even if the dream is forgotten. This indeed turns out to be the case.

There are at least three ways that we learn about human dreaming besides the telling of the dream. We act out some dreams in our sleep. Our brain activity changes. And dreaming helps us learn.

First, sometimes the normal inhibition of large-muscle movement during sleep breaks down, and people, for example, sleepwalk. In this state, asleep but not inactive, sleeping dreamers may act out their dreams. Such enactments are reported by the vast majority of people asked. In such cases, violent sleep behaviors accompany violent dreams. Postpartum women may dream of looking for their baby, and feel around the bed in their sleep. Those who talk in their sleep, if awoken, often describe a dream related to their speech.

These sleep behaviors thus sometimes reflect dream experiences. This is true whether the dreamer can tell of the dream or not. Dreams often are hard to recall, in part because dream experience and dream recall can function independently of each other. One symptom of a sleep disorder in people is the increased muscle activity that allows dreams to be acted out. Patients with REM sleep behavior disorder may act out dreams that they don't recall. That is, patients will enact a dream while asleep, but do not *recall* a dream on waking. It is reasonable to conclude these patients were still *experiencing* the dream during their scenic movements while asleep, but did not retain a memory of the dream.

The second way to learn about dreams without telling them is to look at brain activity. There are patterns of activity in the brain that accompany dreaming, associated both with a specific brain region and with particular patterns of neural activity. Dreaming stops in patients with damage to this area of the brain, but can be unaffected by damage to other brain areas, even if those damaged areas otherwise affect sleep. Dream experience is predictable from activity in the colorfully named posterior hot zone of the brain. This strong connection reveals that we often have dreams but don't recall them, as those two functions require activity in different parts of the brain.

Brain region activation also gives some hint of dream content. Activity occurs in the fusiform face area of the brain when dreaming of faces, in the brain area processing spatial relationships when dreaming of particular spatial settings, and in the areas of the brain involved in perceiving motion when dreaming of movement. In principle, detailed scanning of brain activity could reveal when a dream is experienced, something about its content, and whether it might be recalled on awakening. This level of knowledge from brain scanning is difficult to achieve in practice.

Our brains reactivate recent memories during sleep, a process called replay. During the original awake experience, a particular pattern of neural activity occurs, forming a new memory. During sleep, the same neural patterns may be replayed, and will unfold at the same pace as the original experience. This occurs in humans, but has been more extensively studied in birds and rats. People learning new motor skills who also sleepwalk will do the same thing behaviorally—partially reenacting the new skill while sleepwalking. We further reexperience our newly learned memories in dreams. Thus newly formed memories are replayed three ways during sleep that parallel each other: replayed brain activation patterns, replayed sleep behaviors, and replayed experiences in dreams.

The third way to learn about dreams without telling them arises from the effects on learning of the replay of neural patterns in sleep. Zebra Finch songs are not innate; the finches must learn their songs from other Zebra Finches. To do this, they rehearse aloud, improving their imitation as they practice. The same patterns that fire in the birdsong system of the brain while the finches are rehearsing aloud, researchers discovered, *also* replay at the same tempo in the brain while the birds are asleep. The sleeping finches further moved their vocal cords as they had during song production but silently. Finally, the auditory center of the brain responded as though hearing the soundless neural pattern. The birds were silently rehearsing in their sleep, replaying and also "hearing" the same patterns that produce song when awake.

Birds sometimes do vocalize in their sleep. Sleep-singing has been noted since Roman times. If a bird could talk, might we know what they were dreaming? Alex the Grey Parrot learned an extensive use of words in the interactive setting of a long-running behavioral study conducted by Irene Pepperberg. Alex could talk. He understood words in context. He vocalized. He used words to ask for his favorite treats or toys. Alex rehearsed his new conversational skills in monologues, whether or not there was anyone listening, much as children do when they are learning language, and much as the Zebra Finches do when learning songs. His vocalizations were recorded at the start and end of the day when he was alone, times in which he did a lot of rehearsing. There is no video from these recordings, so we do not know that Alex was always awake while he practiced. From a talking bird, trained like the sleep-rehearsing Alex, we might learn whether birds are dreaming of more than their song and their learning.

WHAT OF HEIDI asleep in my living room, arm tips twitching and body pattern displays moving across her body in waves and sudden starts? We are not certain whether she was dreaming, or of what . . . but could we know this? She cannot tell us verbally of her dreams. Obtaining brain activity information from octopuses in salt water is difficult—such data have been reported from only a single octopus.

Body pattern changes are a form of sleep behavior, however. Much like talking in our sleep or sleep-singing, body pattern changes do not involve the deeply relaxed large body muscles that are inhibited during sleep, which usually prevents whole body motion. Octopus body patterns are muscularly controlled. These muscles are small, like those controlling our own faces, fingers, and toes that twitch in REM sleep. Teaching Heidi a new behavior associated with a body pattern change might lead her to replay that pattern in her sleep, revealing dream content. But already, her body pattern may be enacting her dream.

Octopus body pattern displays occur in particular circumstances: octopus displays encountering a predator are different to that when pursuing prey or finding a mate or exploring the reef. Earlier I described the body patterns of an octopus while escaping an attack by a moray eel—much of it blanched white, turning dark in flight. An octopus attacking prey may move from camouflage to passing cloud to web blanched (but not the mantle) to camouflage again. The detailed sequences of these waking patterns are not well studied. Heidi was not able to tell me of her dream, but, possibly, she was able to show me.

This possibility is a speculative promise that we could learn about Heidi's dreams. The pieces are not all in place yet. Just as hearing about a dream is not the same as dreaming the dream, the experience of an animal's dream will always be subjective. Still, behaviors and brain activity that accompany dreams could allow us a little bit into the world of animal dreams.

The brain's reward system is the same system that drives dreaming—it is integral to motivation. This circuit is also known as the wanting or seeking system. It has a large role in foraging behavior. I speculated that Heidi might be dreaming, and if she were, she might be dreaming of catching a crab. During attacks on prey, the octopus employs a series of body patterns, different from the patterns seen when a predator attacks an awake octopus. Sufficient ecological work on waking octopuses, octopus learning experiments, and further sleep studies could reveal whether these body pattern sequences when asleep might be understood as dream-enacting behaviors. This could reveal both octopus nightmares and octopus dreams.

# 4

## Revelation

# Solitary Octopuses

# 17

## Octopus Hungry and Afraid

**Underwater near Gibbon Anchorage, Prince William Sound, Alaska**

The octopus was out on a flat rock shelf among clumps of turf algae when a dark form emerged into view through cloudy water. The octopus crouched down, spiky, mottled, low and inconspicuous.

I was the approaching dark shape, and I saw this slight shift, which revealed the camouflaged animal. Submerged by ten feet of water, the landscape was almost unrecognizable to me, even though I knew it well. This was familiar ground at low tide, distinctive boulders sky-lining the gravel flats, their tops draped with the golden algae of the mid-intertidal depths, their feet in the lower intertidal among the limp and slippery oarweed and elephant ear broad-leaf brown kelps. At high tide, however, every blade of kelp, algae, and seagrass lifted up in the water, turning the jagged rocky but lush landscape of low tide into a rolling forest canopy that faded out of sight in the near distance into the murky waters of the still rising tide. I was gliding through waters over familiar ground, but I was not exactly sure where I was.

I watched the octopus for a moment as she watched me to see if I was a threat. Which octopus was she, I wondered, and where was she going? A few hours earlier, when the tides were low, I had been on this site with the octopus team, counting. Now I was curious—was this one of the animals we found earlier in her den? What drove her out of her den as soon as the waters rose? And was she on her way out now, or on her way back?

Her eyes were low out of caution. Her body was flat with arms and web tucked in—except in front, where her arms and web bulged out almost as high as her eyes. I looked at this closely—yes, the web between her arms draped over a shape: the edge of a crab carapace showed through the thin membranes of her tissue.

She had captured a meal and was now headed home. I looked ahead of her—which den had she been in? Pushing forward, I came to a distinctive den and the canopy-covered landscape fell suddenly into place and I knew where I was. An octopus had occupied this distinctive den a few hours earlier at low tide. Now it was empty. When I was out of view and she was no longer afraid to move, I suspected the resident would return, carrying a crab to satisfy the hunger that had driven her from safe haven to find food.

HUNGER DROVE HER OUT, and fear halted her progress. Our emotions—how we feel—determine what we do. Do octopuses really experience these emotions? And how do we know? It is not enough to draw the parallel with ourselves. Other animals *do* have much in common with humans, but not always in straightforward ways.

I recalled once in Serengeti, in Tanzania, stopping my Land Rover at the top of a gully. I enjoyed watching the birds flitting through the underbrush, and I ate my lunch overlooking the water. A shadow emerged from the bush and alerted me to the presence of a leopard. I grabbed my camera to collect some photographs of this rare encounter. The leopard put his front paws up on the bonnet of the vehicle, and I took an image of the leopard's furry countenance regarding me, my camera reflected in the rearview mirror. I grinned broadly, joy clear on my face at my encounter with this very relaxed and non-skittish wild cat.

My grin, however, transformed the moment. Carnivores, and many primates, bare their teeth in a grimace to convey threat. The calm demeanor of the leopard changed. He snarled at me and vanished into the overgrown bank with a flick of its black-spotted tail. Giving human shape to animal sensory and behavioral worlds—anthropomorphism— can lead us astray.

Yet anthropomorphism is not always wrong. Universal ecological needs can lead to similarities in how organisms function. All animals, including humans, also share a deep evolutionary ancestry that leads to many surprising parallels, despite superficial differences. Bare teeth and squinting eyes (a smile) can signal not just aggression but also submission among carnivores and primates, depending on other body language. Among humans, smiling can signal joy, agreement, or tolerance (forms of social agreement or submission), depending on the circumstances.

⸺

DO ANIMALS SHARE SOME FEELINGS, like hunger and fear? These feelings are imperative and evolutionarily ancient. There are perhaps no more basic feelings than the urges to eat and to avoid being eaten. Australian physiologist Derek Denton named these ancient and demanding urges the primordial emotions.

Denton studied thirst, an imperative familiar to all readers. Thirst in mammals arises from internal sensors located in the brain that detect the loss of fluid from the cells, a process that accompanies overall loss of fluid concentration in the blood. Other internal sensors respond to a rise in salt (sodium) concentration, also leading to thirst. Thirst is gratified within a few minutes by drinking. The experience of thirst is shut off by other internal sensors and perceptions that together respond to the taste of water in the mouth, its flow through the esophagus, and the distension of the stomach. Octopuses, of course, live in the water. They drink small amounts continuously, but they drink more after eating, a response probably triggered by stretch receptors in the crop (a chamber of the octopus digestive tract, positioned after the esophagus and before the stomach and intestines).

Denton lumps together a number of imperatives as primordial emotions: thirst, breathlessness (air hunger), food hunger, pain, salt hunger, muscle fatigue, sleepiness, the urge to pass urine, the urge to defecate, sexual orgasm, and the urge to regulate body temperature. What these have in common is that they are homeostatic—they maintain the body in a state of physiological wellness. They also become imperative; that is, they are harder and harder to ignore with time and command an action, whether that be drinking water or taking a bathroom break.

Each of these (and relieving them) either feels good or feels bad. These primordial imperatives drive animals to specific actions: those that will satisfy the sensation or relieve the imperative. They have in common that, like thirst, each is extinguished by a particular set of diverse sensory input that direct our attention to gratify the need.

......................................

I ONCE FOLLOWED A LION and her cub as they walked all night, driven from their own home range whose streams contained no water, to cross the dry and nearly empty short grass plains in the Serengeti. They passed silently through dangerous territories of rival prides. Discovery might result in a long chase, and being caught could lead to fatal injury. During this long trek, they did not deviate from their chosen direction. In the wee hours, they came to a kopje, a rock outcrop emerging through the volcanic soil. Here in a rock crevice was a natural pool of warm and greenish water that had persisted long into the dry season. Both lions drank until their bellies swelled, and then after a brief pause, they walked again many hours back to their home range. Only then did they sleep through the remainder of the warming day.

The perception of thirst drives specific actions—not stalking prey, nor patrolling borders, nor avoiding rivals, nor sleeping—in an effort to coordinate the inner need with the outer environment. The lions were aware of and experienced their thirst, and in so doing were able to bring to bear memory, their cognitive map of space, and some knowledge of their own position in the world to resolve the imperative primordial emotion driving them. This is why primordial emotions, the felt experience of bodily needs and desires, exist.

IN THE SEAS, octopuses move water over their gills to breath by rhythmic expansion and contraction of the mantle. An octopus ventilates faster by up to 50 percent when oxygen intake falls below demand, and ventilates more deeply, expanding its mantle to as much as four times the volume of normal resting ventilations. Faster, deeper ventilations result in more water passing over the gills, a necessary condition to extract the needed oxygen. Such large mantle expansions are the same as motions preparatory to a jet-propelled movement or escape response. Similar changes in the ventilation rates (but not volume) are also associated with defecation, defense, and arousal (for example, from sleep). When an unpleasant chemical is introduced into the water with an octopus, they can cease ventilations for several minutes, clamping their gill slits shut against the irritation.

Octopuses take six to twelve hours to digest a meal, and all octopuses accept an offered meal following thirty-six hours of fasting. After eating, the octopus shows a reduced tendency to attack, a well-established fact given that caretakers can feed aquarium octopuses "to satiation"—that is, until they no longer attack the food. This cannot be surprising. An animal without a motivation to start and to stop feeding would starve to death or burst its gut. This reduced tendency to attack is specific to predatory attack. Octopuses that have fed do not always show a reduced tendency to explore. Although not in need of food, they are still hungry for exploration, still curious and interested in play.

Octopuses also experience pain. When injured on an arm, the algae octopus reacts immediately by inking and jetting to escape the source of damage. But these octopuses also groom the wound, holding the injured arm in the beak for several minutes, much as I might suck my finger momentarily after hitting it with a hammer. Octopuses further contract the injured arm, and curl adjacent arms around the injury to protect it from bumping against anything. These protective behaviors persist at least a day after the injury.

In at least two octopus species (the algae octopus and Bock's pygmy octopus, *Octopus bocki*, both found in Indonesian waters), lidocaine applied to an arm blocks behavioral responses to a pinch near the application site. Lidocaine is a local anesthetic that causes temporary loss of sensation; it is an active ingredient in many topical pain-relief creams such as Bengay. In both these octopuses, the animal does not feel the pinch or pain where the lidocaine was applied because the relevant neural responses also cease. Ethanol works in octopuses as a general anesthetic, leading to loss of responsiveness and eventually to unconsciousness. Another compound, magnesium chloride, has similar effects: it blocks the neural awareness of a pinch when used locally, and causes unresponsiveness and eventually unconsciousness when used generally. In these ways, the octopus response to alcohol or lidocaine or other pain blockers (analgesics) is much like our own: the pain-related behaviors, neural signals, and the perception of pain, all cease.

Octopuses also sleep—all animals do. Octopuses settle down in a favorite spot, typically their den. They curl their arms around themselves, close their eyes (or at least narrow them), and their breathing slows. If deprived of sleep, octopuses have to make it up later, sleeping longer or at a nontypical time of day.

In these cases, the octopuses—short of breath, hungry, sore, sleepy—cope and adjust to universal animal needs in the way of active mobile animals: driven to action by primordial emotions responding to internal sensations. Like the usefulness of its contingent sensory world, this is another way that knowledge of self is a central aspect of the octopus's world. These animals do not confuse their own movements with motion in the world, nor their own sensations with events of the external world. Their sense of self, however shaped, is important to coping with their world.

In a simple but not obvious example, consider the fact that octopuses at times eat other octopuses. I have found remains of one octopus in the den of another, the dead octopus partially consumed. An octopus clearly can choose to grab another octopus or octopus parts. They have no trouble grabbing almost anything, including octopus skin, and their suckers adhere automatically. Octopus neural architecture is decentralized, with limited communication of details between brain and arms. Given the notion that an octopus's arm may act autonomously, how is it that the octopus knows its own arms? That is, how do they know not to grab or eat themselves?

The suckers of detached octopus arms, absent any central brain input, curiously do not adhere at all to octopus skin, although they will adhere to skinned muscle and most other substrates. This provides the first clue to where this self-recognition occurs. Something in the skin inhibits the attachment reflex of the suckers.

The way that octopuses react to a detached arm provides a second clue to this self-recognition. Octopuses grasp the skin of detached arms with their suckers, hold it to their mouths, and treat it as a food item. In contrast, when the detached arm is their own, octopuses contact the arm repeatedly, but the suckers seldom attach. They rarely hold their detached arm in the mouth, but if they do so, the arm is grasped with the suckers only at the amputation site (where no skin covers the flesh), and then hold only with the beak, their own suckers avoiding it. The octopuses do not hold their own detached arm as a food item.

Octopus suckers alone recognize and avoid octopus skin in general. Without attention or need to understand where and when their arms are touching each other, octopuses are able to avoid grasping themselves, despite suckers that automatically stick to whatever else they may touch. Intact octopuses can also recognize *their own* skin, and this self-recognition is available for motor control of whether and how to hold on to it.

As these observations show, octopuses respond to their imperious and primordial emotions in ways that require some sense of self. Along with other active animals, octopuses alter their own sensory input as they move in the world. They must keep track of their own sensory organs. They must choose behaviors that satisfy their inner drives and appetites. They must somehow recognize their own limbs, if only to avoid tripping over them or attacking them.

Perhaps octopus emotions and sense of self should be obvious to us. But it is one thing to reflexively assume that, clever as octopuses are, they have complex inner lives. It is something else to defend that scientifically, and to fully absorb that understanding is yet another.

It is not surprising that octopuses experience hunger or other primordial emotions. Whether they feel pain, however, and how that pain can be alleviated, until recently have been open questions in caring for captive octopuses. Many people are surprised to learn that an octopus might dream; and we may not think of octopuses as imaginative. The surprise is visible, for example, when people ask whether an octopus has a sense of self, if they really make choices, and whether an octopus can have a relationship with another animal.

The first motivations, the primordial emotions, are major drivers, carriers of the inner need for action. But primordial emotions, motivating the animal to meet its needs, are not enough for it to do so. Active animals must also sort through the opportunities afforded from the environment and decide how to take action. Their awareness is how these inner needs, which impel action, meet the choices animals make in the face of opportunities.

# 18

## Octopus Cannibal

**Bahamas**

A common octopus keeps her eyes high.

She can just see over the coral head, now dead and grown over with algae. Meanwhile, three and then four arms are reaching under, into, and through the available crevices and holes. She wears a body pattern combining elements of both disruptive and mottled guises. Around the eyes, her skin is pale, but a dark eye bar extends downward from each side of the pupil. Her head is dark; her mantle is also dark except for a transverse pale bar. Her contrasted arms and web are pale and reticulated with dark lines. Two Bluehead Wrasse swim by. They are not focused on her, and she ignores them.

On the far side of the overgrown coral rock, her reaching arm touches—what? Something small instantly flees. Her attention now on that arm, she pales and pounces to that side of the coral. Now behind her, she can see the fleeing quarry, and she pushes off the dead coral, jetting after it. The small fleeing animal makes a nimble left turn. The pursuer spreads her arms, one nearest the target whipping out toward it, but she misses.

She turns now also, jetting and reaching in a pounce that covers the distance traveled by the small fleeing form. As she ends her leap, with the quarry right at eye level in front of her, the pale even color vanishes, replaced by mottling and a dramatic black edging of suckers along each arm. The prey inks and jets away. It is a smaller octopus, less than a tenth the size of its opponent, and it is using every octopus escape tactic against its larger kin. The ink cloud is small and is left behind in the chase. The tiny jet-propelled octopus is gaining ground on the larger pursuer that lands among her own disorganized arms and then leaps off the bottom, pale again, after the prey.

She jets up into the water, flinging out the first right arm, and the smaller octopus flees upward. The pursuer's arm, moving almost too fast, brushes along the fleeing octopus, touches small trailing arms, then head, and finally the forward mantle. Somehow, as the two make contact, one small octopus arm is tangled on the suckers of the pursuer. The pursuer now coils her arm, twisting around to grab the prey tenuously by a single one of its tiny arms. Throwing her first left arm over her own first right arm, and tucking the right under toward the mouth, the pursuer is able to wrap the smaller octopus into a billow of web and surround it. She falls back onto the reef. When her arms contact the bottom, she changes in a moment from pale uniform coloration to a camouflaged mix of browns and cream mottles and stripes.

Octopuses eat one another. No wonder they spend their lives alone.

......................................

IN THE OCEAN, the food web is often size-based. The primary producers at the bottom of the food chain are small phytoplankton suspended in the water. Water is about eight hundred times as dense as air, and so it suspends many particles that would fall through air. Floating algae (phytoplankton) are fed on by small animals (zooplankton) that are also suspended in the water. Larger zooplankton feed on smaller ones. Small fish feed on the larger zooplankton. Many species hatch out as plankton even though they will become big as adults, and thus will occupy higher trophic levels as they grow.

Cannibalism is common in the animal kingdom, perhaps particularly so among water-dwelling animals. Fish from early hatching cohorts grow larger than later cohorts do, and this size difference contributes to cannibalism. The younger cohort may also grow to compete with their larger conspecifics for the same resources, so that by eating them, the larger individuals not only gain a meal but may also reduce competition. Although there are many other ecological aspects to cannibalism, for these reasons cannibalism among marine animals is commonly size-based.

In the incident that opened this chapter, the predatory octopus is much larger than its prey. This kind of cannibalism is common among the cephalopods, according to a 2010 publication. All species of the cephalopods are carnivores, hunting or scavenging their prey. Among squids, cannibalism of smaller individuals by larger has been reported commonly in twenty or more species. Squids form schools of individuals nearly uniform in size—schooling with larger conspecifics could be a fatal decision. Cannibalism also appears common among the most studied cuttlefish species.

I found a beak from a giant Pacific octopus in the midden of another. Once I found a nearly whole but partially eaten carcass of an octopus drawn into an octopus den. Of course, in these cases, the occupant of the den may have found the remains of a dead octopus and drawn it into the den as a meal—it was not possible to discern the cause of death of these remains.

A few individuals of the Maori octopus (*Macroctopus maorum*) captured in a New Zealand study contained octopus eggs of the same species in their guts. When held together in aquariums, larger Maori octopuses attacked and attempted to eat the smaller ones, and also attacked a smaller species, the gloomy octopus, when the two were housed together. While no successful depredation of one octopus by another occurred in this study (even in the confined conditions of an aquarium), the guts of 8 percent of wild Maori octopuses contained the remains of their own species. Although relatively few individuals of this population ate their conspecifics, another octopus as prey can represent a large meal: by weight, octopuses comprise the largest part of the diet of the Maori octopus. Cannibalism by the Patagonia octopus (*Octopus tehuelchus*) and the southern red octopus (*Enteroctopus megalocyathus*) may be similarly common.

........................................

THIS CANNIBALISM IS SIZE-BASED—larger individuals attack smaller prey—but there is also sexual cannibalism among octopuses, in which the female attacks and consumes the male. While observing the day octopus in their habitat during a foraging study in Palau, a team of scuba divers recorded a small male initiate mating with a larger female (about twice his mass). While at arm's distance from the female, the male extended his third right arm, the arm tip entering her mantle at the gill. The third right arm in males is the hectocotylized arm. This modified arm includes a groove along its length and ends in an unsuckered tip; these together aid the male in passing sperm packets down the arm and placing them in the female oviduct. Male day octopuses often adopt a camouflaged body pattern during mating attempts, their skin spikey, body and arms speckled with white spots that blend with the coral rubble of the background. They lay their second right arm across the top of the reaching third right arm during the attempt.

During this observed interaction in Palau, the hungry female continued to forage during this mating attempt, which broke off when the male startled at her movements. Smaller males, even while interested in mating, act leery of their proximity to females. Over the next three hours, the male made a dozen more mating attempts. Most lasted a minute or two, but one attempt lasted fourteen minutes; another twenty-three minutes. Five minutes into one of the later matings, the female broke it off to chase another small octopus, probably male, around the reef for about twenty minutes. The small chased male fled, inking. The female pursued, reaching toward the fleeing male. But this female missed as the octopus jetted toward the surface, and after missing, she broke off pursuit. The fleeing male lived.

The original male made two more mating attempts, but they were brief. Barely more than one minute into the second attempt, the female approached the male, knocked him off a small coral ledge and then pounced, engulfing him in her web and arms. The male inked, but was unable to escape. After moving to a nearby coral cavity for half an hour, during which the male was fully subdued and probably killed, she then returned to her den, where she fed on him for the next twenty-four hours. The following morning, another small male cautiously approached her den, and, never coming into view of the female, reached around the coral to mate. The small male and the sheltered and well-fed female remained mated this way for three hours without ever seeing each other.

ENCOUNTERS WITH ANOTHER OCTOPUS always appear to be dangerous affairs, despite being necessary for mating. In at least a few species including the wonderpus octopus (*Wunderpus photogenicus*), the day octopus, and gloomy octopus, the attacking octopus attempts to restrain another octopus with a throttlehold. The aggressor attacks the other octopus, forming a loop with one arm around the mantle near the gill slits. In these species, as in a great many octopus species, the second arm is longer than the first, and these octopuses usually attempt constriction with the second right arm. The aggressor then tightens this loop, gradually constricting the victim. These attempts do not always succeed, but once this constricting grip takes hold, the captured octopus turns pale, likely a sign of lack of oxygen (although pale body patterns also are expressed during attack on prey, threat response, and in sleep). The constriction prevents the victim from drawing water over the gills, leading to asphyxia and eventually death.

There is not always a size difference between males and females in mating octopuses, but larger females are likely to lay more eggs, and hence may have higher reproductive value. Males are interested in mating with larger females when they find them, despite the risk. The smaller size of males attacked by females makes sexual cannibalism hard to distinguish from size-based cannibalism. The account from Palau of the female eating the smaller male suggests that factors such as male suitability as a mate, and female hunger levels, in addition to size differences, may also figure into when octopus cannibalism occurs. Often the predated individual is half the size or less of the predatory octopus.

Cannibalism by octopuses, and especially sexual cannibalism, is a feature of their reputation as solitary and asocial animals. They are found in isolation rather than in groups. Larger ones will attack and eat smaller ones when they find them. When brought together by a mutual interest in mating, the larger octopus may attack and eat the other.

In a few situations, however, octopuses occur with or interact with others in ways that counter this pervasive notoriety of the solitary and asocial octopus. Some animals of other species share dens with octopuses; waders, snorkelers, and scuba divers sometimes feel befriended by them. Do octopuses in the wild have a sense of themselves in relation to others, despite their being aggressive predators, and against their cannibalistic tendencies? Do octopuses form relationships?

# 19

## Octopuses in Wild Relationships

**Underwater, South of Sydney, Australia**

The gloomy octopus sat in her den, excavated under the edge of an unrecognizable lump of metal jetsam heavily corroded and encrusted with marine growth. From here, she could look out on the passing fish, watch for approaching dangers, and perhaps even see a possible meal moving past. Next to her in the same den were three fish—Southern Bastard Codlings. The codlings, smooth red fish with orange lips and chin barbells, seek and feed on small bottom crustaceans. The three fish pressed against each other, or nearly so, each facing out of the den alongside the octopus. There was hardly more space between the octopus and the nearest fish than between the fish themselves—they were touching at times. The four looked very much at home together, like long-familiar roomies.

The octopus actively maintained her den space, excavating it to prevent the cavity from filling in by waves or storms. She may have to defend the space against intruders. The codlings, as far as my colleagues and I know, do not contribute in similar ways.

The fish benefit from the shelter that the octopus built. But what did they bring to the household? Possibly nothing. Perhaps the fish do not bother the octopus or are too dogged to evict. Having them there may cost the octopus nothing. This may be the case, although the close contact suggests the octopus could quite easily make a meal of the closest of these. Why doesn't she? Perhaps the codlings help the octopus keep the den free of pests since they feed on such small crustaceans. However, the octopus periodically cleans the den herself. The large fish seem to be unlikely, cumbersome predators within the den.

Might the octopus just like the company? Can some invertebrate animals such as octopuses form relationships with other individuals?

ANY ACTIVE AND visually sophisticated animals, including the particularly versatile and flexible octopuses, will have capacities for forming relationships between individuals, whether of their own species or another. A relationship is the stance or attitude with which two or more connected beings regard each other. There is little reason to expect capacities for relationships to exist only among the vertebrates. Instead, we might expect them among all animals that actively move through and manipulate their environments. Should we then expect that octopuses form relationships with others, including perhaps other wild animals?

Octopuses in aquariums certainly seem to do so. The expressive color changes of octopuses and their habit of shooting jets of water at disliked aquarists make this visible. Caretakers readily recognize whether a particular octopus likes or dislikes them. Octopuses *recognize* their caretakers, but they also recognize other octopuses, fishes, and predators or prey, as agents—different from other moving objects like wafting kelp or debris drifting by.

Agents are actors in the environment. Agents fall into different categories, some dangerous, some tasty, others neutral or uncertain. The different categories of agents require specific actions. By exhibiting the correct behaviors, octopuses are able to make their way in their watery realms. These behaviors show their ability to categorize their environment and the agents in it. Many octopuses, for example, will engage with a curious diver in ways they do not with an investigating sea lion. In a test with day octopuses, rather than show fear, some reached out to touch a pencil held toward them by investigating divers. When investigated by a sea lion, a wise octopus will retreat farther into its shelter.

Categorization is strongest for evolutionarily important objects. In humans, specific areas of the brain are dedicated to the categorization of human faces, to animals, fruits and vegetables, and to useful objects like scrapers and knives. Evolution shapes the same capabilities in other animals.

THE DISCERNMENT OF OCTOPUSES was made clear to me one day while diving in Jervis Bay, Australia, where the octopus kept home with the Southern Bastard Codlings. Parts of this bay lie within Booderee National Park; the Koori people of the Dhurga-speaking Yuin nation are traditional owners of these areas.

*Booderee* translates as "plenty of fish." This was true on the day of my dive. As part of a dive team, I was setting out cameras, mounted on small tripods and weighted to stay on the sea floor. As we neared the bottom, the cameras hanging deadweight from our gear, we descended through schools of hundreds of fish, each fish a bit larger than a hand span. Greenback Horse Mackerel flashed silver-sided and yellow in the underwater light, swimming along with dozens of black and white striped Mado with yellow tails. Through these swam Ocean Leather-jackets that usually ignored us, although they can be aggressive in some cases. We wanted to record octopus behavior. On the shell-strewn ocean floor, we settled the cameras down among the small black-and-white Eastern Fortescues and the Blacksaddle and Bluespotted goatfish, which traveled together in small gangs and touched everything with the pair of barbels hanging below their mouths.

Of particular note on the sea floor was a mound, six feet or more in length, of floral milky spots amid browns and blacks, fringed at one end with barbels branching rootlike, and at the other end, the distinctive two dorsal fins and the tail of a shark. This was a large Gulf Wobbegong, whose small eyes hid among its camouflaged body patterns. The barbels fringed a wide forward-placed and downward-facing mouth and broke up the edges of the animal, further camouflaging its head. The shark lay motionless at the edge of our study site. Knowing these animals are lunging predators, we cautiously never swam in front of it as we placed cameras around the site, instead passing several feet over it to accomplish our goals. Then we left, the cameras recording our data.

We returned hours later with replacement cameras, as the battery life on the first set ran low, and repeated our tasks. In the intervening surface interval, I had read that, although wobbegong sharks rarely attack divers, on occasions when a diver had foolishly ventured too close, the wobbegong would clamp onto an arm or a leg. Although their teeth inflicted some damage, these are not the sharp severing teeth of great white or other familiar sharks. The wobbegong, however, holds tight and does not let go, creating a life-threatening situation for such an unfortunate diver. Our team was again scrupulously careful to go above the shark, avoiding passing directly in front of its wide and muscular mouth. Collecting the spent cameras amid the throngs of fish life, we returned to the surface once more.

Here, I quickly reviewed our video recordings, looking for signs of interesting behavior from octopus dens. It was notable, however, how little was going on. At each den, the octopus stayed out of sight for the entire recording. Only one remarkable moment appeared on the morning video. A school of Greenback Horse Mackerel passed back and forth over the motionless wobbegong, staying well above it much as the other divers and I had. Without warning, the wobbegong lunged suddenly *upward* and grabbed an unfortunate mackerel. So fast was the upward lunge and capture that it happened almost between video frames. I showed my colleagues this exciting moment immediately. On our final dive to retrieve video cameras as batteries and daylight faded, *no one* swam over the motionless wobbegong again. We now instead gave it a wide berth of several body lengths around it while collecting our gear.

More interesting still was the reaction of the octopuses. Frequent activity from a den is usual when a wobbegong is not present. An octopus will sit on the lip of the entrance, groom, maintain a den by moving shells, carry the remains of a meal a few body lengths away to dispose of it, and come and go to forage or do other business. In the presence of the wobbegong, however, all activity ceased. Octopuses did not sit at the entrance, nor come and go, nor otherwise make themselves visible.

There was one exception. In the late afternoon as daylight hung on the cusp of the change to submerged twilight, a gloomy octopus cautiously crept into view inside its den, emerging between the Southern Bastard Codlings. Fifteen minutes after the first appearance, the octopus had gradually crept forward and was sitting in the mouth of its den. The wobbegong noticed. It crept closer to the den, as though merely shifting its body in place. It took nine minutes for the shark to move one body length nearer the octopus. Despite its subterfuge, the octopus noticed the move, and its eyes raised high on alert. The wobbegong lifted cumbersomely off the bottom to twist directly toward the den, and it landed its head and mouth less than half a meter in front of the den opening. During this motion, the octopus retreated completely into its den, and was no longer visible. Half a minute later, the wobbegong returned to its previous position on the edge of the site. There was no other octopus activity that day and the next, during which the wobbegong was always present.

THIS EXTREME CAUTION around a fast-lunging sit-and-wait predator seems wise, but it is also a revealing behavior. Octopuses do not react in the same way to every predator. An octopus may freeze briefly or adopt a more camouflaged body pattern when a fast-moving predator looms into view, but once the predator has passed normal activity returns. Swarming predators, such as the sometimes aggressive Ocean Leatherjackets, are usually ignored if near the den, despite the fact that they can mob and kill an octopus that finds itself without shelter. On one occasion, we recorded an Ocean Leatherjacket that took a bite out of an active and conspicuous octopus. The octopus flinched at the wounding, and immediately afterward spent more time in its den than it had previously, but within minutes resumed normal activity. Only the presence of a particular kind of predator shut down octopuses for the entire day—a sit-and-wait predator that lurked motionless nearby, awaiting a careless move.

Octopuses are similarly discerning with their prey. An octopus does not lunge at a swimming scallop, but reaches out almost lazily to grab it. The scallop has limited agency in directing its escape. The same octopus, however, will stalk a fish, and pounce in attack. The fish, of course, is adept at directing its own escape, particularly if it can see or feel the approach. The octopus is reacting to different categories of predators and prey—one is fast and agile, another is slow and clumsy, yet another is a passing threat that can soon be ignored, while others must lurk in memory as they lurk nearby, lest a careless moment result in death.

The capacity to categorize has primordial roots, and itself can be a foundation for sophisticated discrimination. The octopuses are well equipped to make these discriminations. They are active and curious. Using flexible arms and suckers they have diverse ways of acting on the world, such as pouncing, pulling and pushing, holding and releasing. Their discriminations are supported by impressive sensory abilities including recognizing motion, form, and light polarization, the touch-taste of their suckers, and a sensitivity to water movements from pressure sensors in the skin.

When an octopus categorizes its biological environment, the octopus also identifies its role. In one context, the octopus is a lethal predator, searching and pouncing, or taking other predatory actions. In another context, the octopus itself is stalked, and hides, inks, or wrestles as dire circumstances demand. These fundamental roles are ubiquitous in the biological world, but octopuses adopt other roles as well, evoking different sets of responses. Another octopus might be possible prey or predator, but also a potential mate. Some fish are neither predator to evade nor prey to attack.

Octopuses notice these fish as agents in the environment. They tolerate attention from cleaning gobies, for example, even allowing these slender fish to enter the siphon. The day octopus can also tell when a fish is referring to something—that is, pointing, fish-style.

Groupers and coral trout are large, fast predators in the Red Sea, but they cannot follow their prey into the tight hiding spaces available on coral reefs. Other predators can, including, for example, moray eels and Napoleon wrasses. The eels are flexible and can get into crevices. Wrasses have strong jaws that can break the coral, or suck out some hidden prey. The groupers sometimes draw the attention of these fishes, performing a conspicuous shimmy of the entire body in front of the moray or wrasse they are recruiting. The moray or wrasse then accompanies the grouper in a search for prey. If the grouper chases a fish, which then escapes to hide in the reef, the grouper may display a headstand over the spot where the prey is hiding, indicating its location. In this posture, the grouper shakes its head, checks if the wrasse or moray notices, and shakes its head again. It is pointing—to indicate, *"Here is food I cannot get at."*

In Australian waters, the coral trout behaves the same way, but toward day octopuses. Coral trout are drawn to foraging octopuses, attending their searching on the reef, perhaps to catch any small fleeing prey that escape the octopus, or to pick at shells it dropped. On my first diving trip to Australia, I was excited to find the first day octopus I had ever encountered. The water was clear, the octopus was actively exploring the corals, and during the encounter, I focused on taking photos of this new octopus species. Only afterward while reviewing the images did I realize that in every exposure I had also captured her fish friend, a coral trout that in this encounter was by the octopus's side throughout.

The coral trout may also do a headstand over a crevice to which some quarry has escaped and is now out of sight of the octopus. The octopus will then approach the displaying trout and explore the indicated reef crevices. The coral trout do not make this gesture when no octopus is nearby. It is perhaps not surprising that an octopus is as smart as the clever morays and Napoleon wrasses. These animals understand and use the referential nature of these behaviors, indicating to another animal a particular position or object where prey may lay. They are cooperating in their foraging endeavors.

Pointing of some sort is not widely found among animals, but it does occur in a few species. Apart from humans, the great apes and ravens also point with referential gestures. Dogs also understand pointing by humans; dolphins can point out something to each other in this way. The great apes, ravens, dolphins, and dogs are all animals renowned for their social lives, but below the waves, the octopuses and fishes are pointing across the species barrier, connecting not to family or flock mates, but to a found companion, agreeable company for a day foraging on the reef.

THE WAVES CRASHED outside the reef off Mo'orea in French Polynesia, at the mouth of Opunohu Bay. The swell was not huge, but forty feet down, water surged through the grooves between the spurs perpendicular to the reef face. Just a bit farther out, I swam over a coral rubble expanse, following an octopus on a foraging trip. I hung back, letting the animal explore without interference, but then he moved out of my sight around a coral bommie.

As I rounded that edge to look for him, surge from the waves above pushed me forward, driving me close to the octopus, who took evasive action. He moved himself to interpose a bit of coral rock between him and me. Front arms first, then head and eyes, then even the rear arms, he removed from my view. Would he crawl into a hole and end the encounter?

Instead, just ahead of me the octopus raised his eyes above the edge of the rock. He saw me, and I saw him. Just his eyes though; he pulled the rest of his body behind the rock. In this seemingly simple act, the octopus hid its body from me but kept me in sight to observe me. If I were on the other side of the rock, the octopus would be in plain view. To hide from me, the octopus acted on my point of view.

············································

IN ANOTHER DIVING ENCOUNTER, this one in Alaska, an octopus looked at me and I looked at her. She clung to the kelp, camouflaged brown and kelpy. What was in her mind as she gazed back at me?

I was out collecting octopus that day. So, keeping my eyes on her, I reached for the collecting bag clipped to my dive belt. I fumbled. Reached again. Fumbled again. For a brief second, I glanced down to find the bag. I put my hand where it needed to be. I glanced back up to behold empty kelp, ink hanging in the water. She was gone.

············································

OCTOPUSES ATTEND TO the eyes of those watching them. In the kelp, one octopus waited until I averted my eyes before fleeing. In Mo'orea, the octopus positioned itself based on what I could see. Octopuses show their awareness of us, their understanding that we may be important actors in the octopuses' environment who carry intentions or goals that could matter. To act accordingly, the octopus acts in relation to the action we might take; the octopus uses something of our perspective; and the octopus acts in relation to us.

These abilities are the foundations supporting basic relationships. We should expect octopuses to have them; indeed, at one level—the relationships of predator to prey, of prey to predator—they are inevitable. Octopuses carry this further, relating to the intentions, position, and attention of others in complex ways. Despite being solitary animals, octopuses could be in relationships with others. If so, do octopuses ever gather? What happens then?

# Society Octopuses

# 20

## Gathering Octopuses

**Underwater, South of Sydney, Australia**

The octopuses certainly saw me coming. I descended about sixty feet to a sandy plain interrupted with clumps of algae and live scallops. From above the seafloor I saw a patch that stood out from the rest—a pile of thousands of scallop shells accumulated over time. As I headed in that direction, from this shell bed several octopuses noted my approach.

Octopuses usually live solitary lives, including most individuals of this species—the gloomy octopus. But here, at times as many as sixteen octopuses can be found in close quarters—the span of the accumulation of scallop shells is no more than would fit in a modest suburban living room. Scallops are the main food of these octopuses, and the emptied shells have accumulated around a single human-made metal object, probably dropped from a boat. The object is about as long as my forearm, and is now largely buried, but octopuses still excavate one or two dens along its edges. The shells of their meals are scattered at the mouth of the den, where they stabilize the silt, making it easier for them to create a place to live.

This is Octopolis. Some years back, local diver Matt Lawrence discovered a collection of octopuses at this site. He and colleagues now visit the site a few times a year, to learn more about the octopuses that live here. The area can be very busy in the Australian summer, and nearly deserted in the winter. There have been octopuses here over many years. However, as gloomy octopuses do not live more than a year or two, the octopuses we find here one season are not the same ones we return to the next.

A small octopus in a den along the edge of the site extends herself out toward the clumps of algae and live scallops. She leaves her den. After moving into the nearby algae, she pauses. She sits, reddish-brown amidst green clumps, but in the ocean-filtered light, the contrast is muted. Her skin is scattered with raised papillae that break up her outline to blend in with the algal fronds. Once motionless, she is hard to pick out from the background. After a few moments, she resumes her travels in a low crouch, as high as the algae, until she is about ten feet from her den.

The live scallops close by the shell bed are small and sparse, because the octopuses have been taking the bigger individual scallops to carry back to their dens to eat. Scallops farther out are two to three times as abundant, and often twice as large, but the farther she goes, the riskier it is. And while it's not exactly safe here, her den is quicker to reach if she has to flee than if she was out in the distance where the big scallops are found.

She covers a clump of scallops and her body jerks slightly in a tugging motion as she pulls the biggest ones free of their attachments to any bit of hard shell half-embedded in the silt. Then, carrying the scallops under her web in front, she hastens back to her den in a rapid walk. Once in her den, she sits for twenty minutes, consuming her meal.

Her den is excavated into a continuous bed of discarded scallop shells, the sides of the burrow stabilized with them. Some were added to the accumulation days ago, and were covered with a film of diatoms. Others have been here for months or years, now overgrown with encrusting worms or barnacles, bryozoans, or the eggs of marine snails. Tiny hermit crabs, beautiful pinstriped snails, and small fortescues and stonefish, even a juvenile blue-lined octopus (*Hapalochlaena fasciata*) crawl and live in the shell pile. The bite of a blue-lined octopus can kill a person, so we are careful not to touch this tiny animal. It is smaller than my thumb and content to crawl under a shell to hide.

Just a few feet away, the small gloomy octopus pushes the cleaned scallop remains out from under her web. She is done eating. The shells cascade onto the low mound of discards surrounding the mouth of her den. The nutrients left from the tissue of her meal will nourish a bloom of diatoms or other algae on the pearly inside of the shell. The accumulation creates a shifting but complex habitat for the small encrusting organisms and other animals that come to graze the algae or prey on the worms or barnacles.

This shell bed also draws a surprising diversity of fish. In addition to the fortescues and stonefish, a Dwarf Lionfish wedges along an encrusted ridge of the central artifact, its red-banded dorsal spines erect and needle sharp. Banjo Sharks, like mats across the seafloor, often lie on top of one another, so still that they themselves might be scattered with hermit crabs. Tan and dapper-painted Port Jackson Sharks with black eye masks sit among them, and a few broad Kapala Stingarees rest here and there. A cow-eyed porcupinefish, spines flat along her sides, comes briefly to look at the just-discarded shells, her fins paddling wildly as she maneuvers. She wears a look of perpetual joyous surprise, bony mouth pursed and eyes wide under brows drawn in by arched yellow spines.

A school of Blacksaddle Goatfish with ochre stripes and a black spot on the base of each tail briefly follow the edge of the shell pile, the lead few fish pausing a moment to investigate the silt while the trailing individuals pass over them and take the lead. A coffee-and-cream Halfbanded Seaperch attends them for a moment, and then veers away. Roaming individually in the water just over the area are brown-striped Ocean Leatherjackets with yellow-tinged mouths, while a few mirror-sided Skipjack Trevally and hundreds of silvered mackerel swim around and over the site in a dense swarm that obscures the view as they pass.

The seabed everywhere around is loose silt or sand at least a foot deep with no solid surfaces, covered in this one spot by discarded scallop shells surrounding the central artifact. At some distance but less than a kilometer away, is another area with hard substrate that divers Martin Hing and Kylie Brown would discover a few years later, where bedrock emerges from the sediment in three small distinct patches. The biggest bit of exposed rock there is only a meter and a half across, and the entire area is less than twenty meters in extent. At this second site, too, we found octopuses living along the edges of the emerged rock, where they had also accumulated the shell hash of discarded prey remains. Within meters of this spot, there were rich beds of doughboy scallops and mud arks (another type of bivalve).

The number of octopuses at these two sites suggests octopuses are drawn to them. The octopuses live within sight of one another, and are keenly interested in one another. The abundance of food within a few meters of each site makes them desirable. The octopuses are not going far from their dens to get each meal.

Octopuses routinely carry their food back to the den to eat, and discard empty shells at the den after the meal. Perhaps the mound around a den can get too high over time. We recorded some octopuses carrying the remains away from their dens to the edge of the shell hash, and dumping it there. More shell hash means room for more octopuses, and more octopuses amass yet more discarded prey remains.

The loose silt and sand all around are grown with algae and make good habitat for scallops and mud arks, but silt and sand and mud allow little shelter for these octopuses, who den in reefs and under rocks. Many predators and dangers swim through these waters. In addition to the wobbegong sharks and the Ocean Leatherjackets, we have also seen fur seals, dolphins, smooth hound sharks, and even a penguin in our recordings. Shelter is a requirement for these octopuses. Octopolis appears to have been seeded by at least the artifact, whatever it is, which is clearly artificial. The second site, however, where bare rock emerges from the silt, and which we have nicknamed Octlantis, is entirely natural.

In these waters, where food is abundant, predators common, and suitable shelter only in rare patches, an octopus cannot be choosy. Using the only available shelter, they come into close proximity with one another. This, as it turns out, forces them to deal with each other. They do not turn to cannibalism, however—at least not as a common outcome, which is perhaps surprising.

There are a few other situations where octopuses engage with each other. In the late 1970s and early 1980s, the Larger Pacific Striped Octopus came to scientific attention. Accounts of this animal were unlike any other—these octopuses reportedly formed colonies of thirty to forty individuals, shared dens in mated pairs, mated beak-to-beak, and spawned over extended periods, some eggs hatching as the female was still laying and tending more. Could this be?—in a group of organisms known for their cannibalism, cautious and distant mating, and death after tending eggs? However, a full scientific description of these behaviors was rejected, and not resubmitted. That account was never published.

The Larger Pacific Striped Octopus was undescribed, without a scientific name. It was a striking species, about a hand span in size and one of a few species of harlequin octopuses, distinguished by semipermanent stripes and spots visible among nearly all body patterns. The black and white Larger Pacific Striped Octopus had transverse bands on the mantle, and polka dots in dense profusion across the arms and web. This default high-contrast pattern faded in certain body displays, but left behind thin white mantle lines and white spots on bumps of granular skin of the arms and web.

Critically, this octopus species disappeared after the early 1980s and was no longer found to study. In the 1970s, its known range was small. It occurred in the Bay of Panama and along the northern Pacific coast of Columbia where shrimp trawlers caught it. Rare individuals turned up in Guatemala and Mexico. It became almost legendary: there was next to nothing about it in the scientific record, but there was still an occasional rumor of a lost octopus. Had the species gone extinct?

There the situation rested for two-and-a-half decades. There had been extreme and unusual claims made of the social nature of this octopus, unpublished in a traditional behavioral account, but in an abstract or an aside in other reports. No one knew how to find further wild populations, or even if they still existed.

Then, in the summer of 2012, a few octopuses matching the description of Larger Pacific Striped Octopuses showed up on the aquarium trade. Divers licensed to collect and export octopuses for aquariums collected them all at one location in Nicaragua. The Steinhart Aquarium in San Francisco acquired many of these octopuses. For a short time, they were displayed to the public. Roy Caldwell with other scientists published an account of their behavior in 2015. By then, the octopuses were no longer available to the aquarium trade. Despite interest from the Steinhart, the exporters could find no further supply.

The divers collected all these octopuses from a single aggregation that persisted in the wild for at least two years. The detailed study of behavior in captivity also matched the earlier observations from the 1970s. Octopus pairs would share a den at times, although more often they denned separately. One mated pair even shared food, holding between them a shrimp the female was fed, as they ate together beak-to-beak. Matings also occurred in this position. Females accepted matings from multiple males. Two other housing combinations were tried (male-female-female and female-female, but not male-male); the first appeared to be acceptable to the octopuses, but the female-female cohabitation was discontinued when one female began to eat the eggs she had previously laid. Continuous egg laying is known in only one other octopus species, the closely related lesser Pacific striped octopus (*Octopus chierchiae*). Other species lay their eggs over a period of a few days, tend them, and then die about the time the eggs hatch. The Larger Pacific Striped Octopus females, however, had a long period of egg laying, and later hatching, each continuing daily over three to six months.

There is one other octopus species (*Octopus laqueus*), found in Japan, where male and female mated pairs also share their den. They live in close proximity in the wild, and males and females in captivity will tolerate sharing a den in preference to going without, but would rather have their own den. In short, they balance the need for shelter against the desire to avoid others. This is more tolerant than many octopuses, and it may be the same balance sought by the gloomy octopuses in settling at Octopolis and Octlantis. *Octopus laqueus* was first described in 2005, and as yet there is no published English-language detailed account of its behavior in the wild.

The previously mentioned algae octopus is another species involved in common encounters with each other. This is the long-armed octopus that can drop part of its arm when attacked. These occur in some of the same waters (stretching southwest from Japan toward Taiwan) as *Octopus laqueus*. They are active during the day in intertidal shallows where larger females maintain dens that they surround with collections of small pebbles. Males interested in mating occupy dens adjacent to these females, to be within arm's reach and able to mate with them. Males, however, collect many fewer pebbles. Dens are clumped spatially, and males on the move encounter males in their dens guarding a female. This leads to some aggression. The winner is usually the larger male. Octopuses also encounter members of the opposite sex, and these encounters end in mating, or ignoring one another, or, in the extreme, cannibalism, where the larger female eats the smaller male. In other words, male algae octopus seeking mating opportunities face high risks of aggression from other octopuses.

### Monterey Submarine Canyon, off Southern California

ON A VERTICAL FACE of a canyon wall close to a mile down in the ocean, high currents expose bare rock. Such hard exposed habitats are rare in the deep sea, much as they are in the silt and scallop habitats of the gloomy octopus. On this rock outcrop, a female octopus (*Graneledone boreopacifica*) tended her eggs for nearly four and a half years. A remotely operated vehicle first filmed her in April of 2007, and visited repeatedly until October 2011. Octopus mothers, to continuously tend their eggs, do not hunt or feed during their egg-brooding period, and this female was no exception. She ignored nearby prey and refused food offered by the ROV operators. The female and her developing eggs were still there in September 2011 but she was gone in October, with just the tattered cases of her hatched eggs to mark the spot. Tropical octopuses may brood their eggs for a few months, but in colder water, it takes longer. The giant Pacific octopus, a cold-water species, broods for about six months. The deep sea is cold indeed, and this octopus holds the record for the longest brooding period known for any animal. Egg-tending octopuses and fish aggregate where hard surfaces are exposed, due to the rarity of such habitat. In one such area, researchers counted over two hundred octopuses in a single survey, at times close enough that one photograph captured at least eight octopuses.

The substrate can be very important to octopuses. When appropriate shelter, hard substrate, mating opportunities, or brooding sites are scarce, different species of octopuses must tolerate being near each other. In some cases, there may be enough space that brooding octopuses do not interact much with each other. This appears to be the case on the deep-sea canyon walls. However, on nearby Davidson Seamount, nearly three thousand female *Muusoctopus robustus* mothers clustered their nests around warm water seeps at 3,200 meters deep, each egg-tending octopus within easy reach of neighbors. In other octopus species, pursuit of mating opportunities means that octopuses must contend with rival males or cannibalistic mates. At Octopolis, in the dangerous waters south of Sydney, octopuses returning with prey to dine safely in their dens have created new habitat, and now must contend with new neighbors who moved in.

This proximity could pose challenges for octopuses to the extent they have evolved to be solitary and lack adaptations to cope with other octopuses. Mating and related behaviors do not fall into this gap, nor does cannibalism, but what of anything in between? Can these octopuses relate to each other in any different ways?

## Underwater, South of Sydney, Australia

A GLOOMY OCTOPUS pushed at a shell. A moment later the octopus, a little female, pulled fully into her den, the vertical shaft hollowed into the scallops and silt. Under her arms, she gathered a load of loose shells and debris that had settled from the sides and edges. She rose up, lifting everything to the lip of the den and sat a moment, checking for dangers. Cautiously, she left her den a short distance carrying the debris, and then dropped it all. The octopus then hastened back to her den. Over the next half hour, she repeated the action, then again, conducting three den cleaning trips in all. In her den again, she pushed and arranged a few more shells on the edges.

Were these behaviors automatic reactions to the shells shifting in her den—a sort of automatic or semiautomatic response that unfolds without attention or thought? Or was this octopus thinking about how she would like her den, tidying up, with a more orderly home in mind?

After tidying, she sat on the edge of her den a while, watching the water around her. Fish swam by at frequent intervals. Some, like the Ocean Leatherjackets, are potentially dangerous; others, like the schools of Mados, not so. Then she left her den, traveling farther across open ground than she previously had to dump the debris. When she returned to her den, she was carrying a large black sponge she had picked up, about half the size of her mantle. She dropped down into the safety of her den, holding the sponge on the lip of the den.

Later she retreated into the opening, and with her last arm to enter, pulled the sponge after her. The sponge filled the den entrance forming a door—a barrier to any threat or harassment. The arrangement of objects around her den entrance, the attentive removal of debris, the fetching from a distance of a sponge particularly suited to block the opening— these appear to be intentional behaviors.

Where do mental images—this inner state of knowing a goal— begin? The edges of intention are gradual: there are gradations from inflexible but adaptive actions to behaviors guided by the explicit formulation of a goal. Somewhere along this continuum are motivations that bring about improvements to a living arrangement. And how do these routine abilities relate to the challenges of managing interactions with other octopuses?

Over the half hour this little octopus was den-cleaning, other octopuses in the area were active, as well, close enough that the little female knew of their presence. How might she regard these other octopuses— these predators or mates, or perhaps, neighbors? An odd experiment with a recreational drug revealed a surprising and often hidden capacity within octopus physiology. Octopuses, along with vertebrates, including humans, respond similarly to the recreational drug ecstasy, which increases "prosocial" behaviors such as choosing to spend time with another and touching more frequently. The serotonin and oxytocin hormonal systems that regulate this and other social responses are evolutionarily ancient, and apparently can be active in octopuses despite their solitary and asocial reputation. Oxytocin, in particular, may not simply promote sociality, but instead may clarify social-related sensory perceptions against the sensory background. Such study of octopuses has barely begun, but if these pathways when activated really regulate octopus behavior, this may provide a way to understand why the little octopus not only noted other octopuses in the area, but monitored their activity, and once left her den to interact with another briefly. Certainly, the sponge could serve as a barrier against predators, but was it also a door to close for privacy from socially curious neighbors?

# 21

## Octopus Quidnuncs

**Underwater, South of Sydney, Australia**

There are other gloomy octopuses on this same shell bed. Across the few meters of the site from the den of the little female, on the south edge of Octopolis, another, larger female rose up out of her den and began to move toward the edge. The southern female's movement was noticed, however, and a male octopus, also large and resting a few body lengths away, reached out a long arm toward her. He began to move in her direction. She pulled back into her den.

A few minutes later she tried again; he reached again. She persisted a bit and he approached. She retreated into her den a second time. On the third try, without hesitation, she headed out of her den and directly off-site, toward the sand and algae flats where live scallops abound. He reached again and approached, but this time she ignored him.

........................................

MALE AND FEMALE OCTOPUSES look similar, although there are some differences. Male gloomy octopuses have clusters of three or four enlarged suckers about a third of the way down the second and third arm pairs, which females lack. These are not often visible, but when they are, they anatomically identify a male in a way that we can see on our videos. We can more often distinguish males and females in our videos by their behavior, much as a male Red-winged Blackbird can be identified by his singing atop a cattail. The male octopus mating attempt is distinctive. He reaches with the third right hectocotylized arm and typically covers that arm at the base with the second right arm. Males also are careful of their third right arm, and carry it curled and tucked protectively a bit closer to the body than the other arm tips. Even when not mating, however, male gloomy octopuses here behave quite differently toward females than toward each other.

........................................

AS THE BIG MALE WAS DISTRACTED, another female exited her own den near the center of the shell bed, edging off the site. The male redirected toward her, and trailed this female to the edge of the shell bed. He was reluctant to follow farther.

Minutes later, an octopus approached from off-site, bearing the rewards of a foraging trip. As she first emerged into view in the distance, the large male alerted, raising his eyes and turning dark. Then he stood tall on his arms, mantle behind him rising from a relaxed position. Appearing large and ominous, he approached the arriving octopus cautiously. He reached toward her, while she skirted to one side around him. The male and female touched or reached just shy of contact. The male's caution, his dark skin and tall posture all collapsed at once—he dropped into a walking stance, his skin lightened, and he accompanied the female, ushering her into the southern den previously vacated, where she was already headed in any case. She settled there and began her meal. The male showed himself similarly to the next arrival, who was also bearing a meal of scallops under her web, before she skirted past him directly into the central den.

........................................

THE MALE GAMETE IS SMALL and mobile. The female egg, with yolk for the offspring, is large, and would require more energy to move around. Each sex specializes at this cellular level—one on motility, the other on provision. Just as the roles of being predator or prey are thrust upon organisms by circumstances, so too are the roles of male and female. These roles affect not just the way gametes evolve, but also shape behavior of individuals.

The large male also seemed to have a particular role, one we have seen occupied repeatedly over the years when we visit Octopolis—albeit, given the short lifespan of gloomy octopuses, not by the same individual on each visit. He is the octopus Quidnunc on this patch at this time, and he attends to every octopus that moves. The Quidnunc is the most active individual. In his attention, it is as though he were gathering local gossip, asking, "What now?" when an octopus moves in the den, or "Who is that?" as one returns to the site.

The Quidnunc seldom occupies a den when there are several other octopuses around, and rarely leaves the site, but spends the day crossing and recrossing the shell bed to check and vet each octopus that comes or goes or draws his attention. He is the only octopus who routinely visits dens just after another octopus has left them, feeling inside the empty space briefly before returning to coursing back and forth. This octopus does not behave like the others—not like the females—and if there are other males about, they do not behave like this male, either.

Does the existence of such a role require that octopuses recognize each other as individuals and form relationships? Sorting out recognizable individuals with confidence at this site is fraught, and we are still learning the extent to which octopuses sort out each other in this way.

Distinctive marks or scars allow people to recognize some octopuses as individuals. Both the lesser and Larger Pacific striped octopuses and the wonderpus octopus, for example, are boldly striped species, each of which has individually distinctive patterns of markings. Nevertheless, for most coastal species of octopuses, including the gloomy octopuses, body patterns are nearly all changeable.

At Octopolis, however, we are sometimes able to recognize individuals. My collaborator Peter Godfrey-Smith noted particular distinct and recurring shapes within the changing body pattern displays. A female present in one year had papillae with white tips under the eyes that distinguished her from other octopuses. Another had a unique peanut-shaped frontal white spot. Some octopuses bear distinct scars. In one video, we recorded the moment when the Quidnunc was nipped on the mantle tip by a leatherjacket, leaving an unambiguous white mark. Damage or partial amputation of one or more of an octopus's arms are also common. In another video sequence, we observed that one octopus was regrowing the first right arm from near the base; one, the second right arm; another was healing from injuries to the second, third, and fourth right arms.

We recognized, at best, about one third of the animals present by scars and marks. When distinctive markings were not enough, we also tracked individuals by their movements. In this process, there are good clues but also we must make assumptions. The process is painstaking, as inevitably the octopus leaves the shell bed entirely, completely breaking video continuity. Absent distinctive marks, when an animal then comes on screen from afar, we cannot be certain whether the arrival is of a previously unseen individual, or the return of a familiar one.

Most often, hints included an arriving octopus that came from the direction of a previous departure and headed directly to a previously vacated den, like the southern and central females whose returns from foraging attracted the attention of the male. When we presume this is a returning individual, sometimes we can test the idea with another camera from a different vantage. When we do so, we see that the departing octopus did not go far, was (however distantly) in view the entire time, and the arrival was indeed the same earlier-departed individual. This gives us confidence, but not certainty, that we are tracking the same individuals when limited gaps in their on-camera continuity occur.

This limit on recognition creates a challenge for our understanding of octopus interactions. Is this site, as it seems, an assembly of neighbors who are all familiar with, and come to recognize, each other over a period of days? Or might the octopuses be limited in their ability to recognize each other as individuals, just as we are limited in recognizing them? Perhaps to the octopuses, as well, there is uncertainty in whether an octopus approaching the site is a previous neighbor or a new arrival. It would seem easier for the octopuses if they could recognize an established mate, or a rival that had bested them previously, and to whom they now would be better off to give a wide berth.

Cephalopods do indeed appear to recognize individual people, although much of the evidence is anecdotal. However, in one published aquarium experiment, giant Pacific octopuses recognized individual caretakers who earlier either were nice to them (feeding, whom the octopuses approached even when no food was offered) or were not nice (bringing a brush too close, whom the octopuses would avoid even when the bristly stick was absent).

Amethyst, an octopus in the aquariums at my university, took such a dislike to me, and she would aim jets of cold saltwater at me when I stood close to the tank. One day, visitors were being introduced to Amethyst. Not in the mood for a soaking, when the aquarium was opened I positioned myself with the raised lid (a clear acrylic panel) standing between the octopus and me. As she came up above the waterline where I was presumably in view, Amethyst jetted a mantle-full of icy water, up *over* the barrier, which arched to hit me directly in the face. Once again, I was outsmarted by an octopus. That day, she squirted no one else in the group.

Possibly, octopuses also recognize each other, although that evidence, too, is sparse. In another aquarium study, the common octopus recognized neighbor octopuses and remembered them, at least for a day. This study, and the one of octopuses recognizing their caretakers, provide hints that octopuses recognize each other as particular individuals, but this remains uncertain. Most octopuses may seldom encounter one another in the wild, so that opportunities to exercise this ability might occur in just a few instances in a lifetime.

OBSERVATIONS FROM OCTOPOLIS suggest individual recognition. In one instance, a female and male octopus occupied neighboring dens; the female at the lip of hers watching events, the male out of sight down inside his. In the near distance, a third octopus (a presumed rival male) approached another den. The watching female immediately left her den and grappled with the encroaching rival male before returning to her original den.

The rival male came toward her in that den, and she reached to fend him off. Instead, however, this rival entered the neighboring den, still occupied out of sight by that first male. A tussle ensued. The female reached an arm to get involved, then a second arm, before the original occupant emerged up out of the den, evicted by the rival intruder. The evicted male backed off from the den edge, but the female did not, reaching a third arm and a fourth in after the intruding male, before entering completely into the den, just the tip of her mantle visible in the tangle of arms.

A second wrestling match took place underground in the tight space. A long arm shot out of the den past the female, reaching for a sucker hold, and the intruder rival male pulled himself up into view. He struggled his way out from the den, but the female had him with one or more arms, and he dragged her up from the hole after him as he moved. One of her arms wrapped under his mantle, seeking a gill slit. She attempted the constriction grip, a killing hold in octopus battles.

The rival male dragged himself backward across the shells, pulling the female after him. When she was stretched taut, two arms engaged with him opposite two arms that were pulled at last from the den, she released him, and then scooted back to her original spot. Moments later, the evicted original male also scuttled back into his den. The female offered the returning male a touch on the head with her second left arm and the two tussled a moment before settling down again together as before.

.........................................

IN THIS BRIEF PERIOD, the female saw a male octopus, crossed a portion of the shell bed to wrestle with him, and watched him evict her neighboring male. She then evicted this intruder herself, possibly attempting to throttle him, before welcoming back her neighbor without contest. This account is not conclusive proof of individual recognition, but it is suggestive. The female *did* distinguish the two males and enacted different attitudes toward each. Perhaps she made her discrimination based on the rival males' size, the intensity of some display, or chemical signals. Females of many animal species use each of these to select mates. Individual recognition, where it occurs, can also be part of a female's tool kit in choosing her mates. However, merely making the right choice when interacting is insufficient to demonstrate that animals recognize each other as individuals. We do not yet know the mechanisms that drove this particular complex interaction, nor whether these octopuses make similar decisions often.

It is worth noting that the evidence for cephalopod recognition abilities is weak. Mothers do not recognize their young. There is no parental care after hatching. There is no known kin recognition. Cephalopods recognize their species; they can tell whether others are male or female; they probably make decisions about one another based on size, circumstances, and particular behaviors such as the Quidnunc male's pose of standing tall and turning dark. We still have more to assess with this group of animals. But at Octopolis, at least, octopuses do not seem to follow merely situational strategies, such as "resident stays, occupant leaves" or "the octopus next to me is my mate."

Individual recognition reduces aggression. Recognition can be the basis for a social hierarchy, returning to a mate, or forging alliances. Many species rely on recognizing each other. Some birds recognize their individual neighbors by their voices. Dolphins have signature whistles that identify them individually within their social groups. Asian elephants know each other by smell, touch, and sound. Paper wasps (*Polistes fuscatus*) recognize their nest-mates' faces. Facial recognition among wasps (and primates) is holistic; it is based not on recognition of one or a few distinctive features usually, but instead on the familiarity of the entire face. Recognition is essential to the formation of primate societies, to the pride or pack structure of lions, wolves, and other social carnivores, and to the ability of female King Penguins that, upon returning from foraging trips, find their mates (who are tending the eggs) by knowing their voices, which are calling to them.

The effect of familiarity alone—that sense that "I know this individual"—is less helpful than remembering the context and outcome of past encounters: recognition plus memories. To know is to have access to a bit of information; to remember is to recall a scene, to hold an event in memory. Remembering momentarily brings to mind an individual's own past with details of subject, place, and time. Event memory entails a past moment of self.

Event memory was once considered a uniquely human ability. This notion arose because the retrieved memory was accompanied by the conscious experience of remembering that past moment. Verbal recounting of an event gives us unique access, in a way that is not available to other animals, to understand human abilities to recall past events. Event recall is a broadly defined ability that encompasses scene formation, sequencing of events, location of the self as a viewpoint, and more. Event recall is also loosely specified in that the definition entails consciousness (itself hard to pin down), and in that identifying the precise components of recalling a specific past event has proven to be challenging. What does that mean in terms of animal behavior, or even a person's verbal report?

The key features of this recall are that event memories are associated with a past time (that is, "recently" versus "long ago," or "on my birthday"), with a location, and with what happened. Western scrub jays scatter-hoard hundreds of small food items in the wild and return later to retrieve them. In controlled experiments, scrub jays returned to a particular spot (where) to retrieve insect items (what) only for a short time after stashing while the insects remained palatable (when). If too much time elapsed since stashing a particular insect, the hungry jay avoided returning to that spot, and instead retrieved seeds that do not degrade.

Many animals have memory of this kind, referred to as *episodic-like* memory. That is, the animal has demonstrated retention of what, where, and when, but it remains uncertain that it achieves this performance by recalling the past individual event. Episodic-like memory has been experimentally reported for several mammal species, a few birds, and a fish. In addition, two invertebrates show these aspects of memory: bees and cuttlefish, a relative of octopuses. Common cuttlefish (*Sepia officinalis*) in a controlled experiment were presented with a preferred prey (shrimp) that appeared at fixed intervals at a particular location, and a less preferred prey (crab) in a different location on a more frequent schedule. The cuttlefish searched for the shrimp after long delays since previously feeding, but after short delays, looked where the crab were found. They had learned that the preferred shrimp would not yet be available at that time and place. Cuttlefish thus learned a what-where-when discrimination, suggesting episodic-like memory. This what-where-when memory may not be identical to recalling a specific event, but it could be a behavioral indicator of it, or part of what is necessary to recall an event, a foundation.

Animals that recognize other individuals can use their shared history in making decisions and forming relations. The female octopus choosing between her original male neighbor and the intruding male may have relied on history with each of these individuals, trusting one to be trouble, and the other to be a congenial neighbor. The octopuses in the aquarium experiment, and Amethyst who squirted me, used their history with caretakers to make decisions about how they would interact with us. These octopuses encouraged some caretakers to keep their distance. They readily approached other caretakers—those who, in a word, were trusted. To trust in an aspect of the environment is to learn how it behaves; to have confidence that it will act in a predictable way for better or worse. Trust is built into expectations, and expectations are part of both categorization and recognition.

Do octopuses display individual recognition sufficient for trust?

Do they use memories of a relationship in making decisions? It would behoove the residents of Octopolis, if they could, not to mistake the Quidnunc for another. Other octopuses might be content to live and let live, but the Quidnunc is not one of them, and involves himself with others at every opportunity. The Quidnunc is full of bluster and threat, that when necessary may be backed by physical force.

At least one scholar has likened the relations of octopuses to those of nations. Which brings us to the role that the dynamics of octopuses almost played in international affairs. In the tensest days of the Cuban Missile Crisis, at the height of the Cold War, Soviet Premier Nikita Khrushchev was completing the installation of Soviet nuclear-tipped missiles in Cuba. At the time, the anthropologist Gregory Bateson was studying octopus conflict in aquariums. Bateson had been developing his ideas from these studies, and what he had learned from watching octopuses in close quarters was important.

As the US and the USSR came to the brink of nuclear war, President John F. Kennedy needed to hear what the octopuses were doing.

# 22

## Octopuses in Domestic Relationships

In October 1962, the rules of engagement during the Cold War assumed particular urgency. Soviet nuclear warheads were in Cuba—and ships carrying missiles, launchers, and more warheads were on their way. The US needed to establish new rules to communicate with and relate to the Soviet Union. Gregory Bateson, an interdisciplinary scholar, looked to his own studies with octopuses for insight into this problem.

Bateson understood that for birds and mammals, communication was rooted in parent-offspring bonds. In courtship feeding among many bird species, for example, the courted female begs like a young bird and allows the male to feed her. Bateson recognized that the feeding in this context is a signal, because feeding is not its only function. The behavior's additional function is courtship; that is, relationship building. Courtship feeding is a behavioral metaphor, an implicit comparison between one relationship (parental care) enacted as another (courtship).

In the context of the communications between nations, Bateson looked to another metaphor: that of closeness, physical proximity, which he observed in octopuses. Octopuses were interesting because females tend their eggs, but otherwise lack maternal care of offspring. Octopuses are also notoriously solitary. These facts drew Bateson's attention to their willingness to tolerate the proximity of neighbors as a metaphor for tolerant relationships among nations.

Surprisingly, however, and despite their solitary reputation, octopuses like closeness. Thursday was an octopus that my daughter Laurel and I kept in a home aquarium for a while. Thursday was eager to interact with Laurel. On coming home from school, Laurel would put her fingertips in the water, and Thursday would leave her den at the other end of the tank, scoot along the bottom, and then jet up to the surface for a hello. Even after feeding, she liked to hold on to Laurel, sometimes for as much as a half hour or more. When I chose a seat in the living room to read, Thursday would often quietly relocate in the tank to the point nearest me. She would crawl up and down the glass in my line of sight until I attended to her. By contrast, when Amethyst squirted me with water in the lab, she not only kept me at a distance but also metaphorically indicated her dislike.

The relational communications in octopuses are not rooted in parental care or mating dynamics. This insight allowed Bateson to wonder how the same mechanisms might serve in the relations of nations. Bateson studied juveniles of either (or both) Verill's two-spot octopus (*Octopus bimaculatus*) and the California two-spot octopus (*Octopus bimaculoides*)—he did not always distinguish which. Bateson collected his octopuses along the shores of La Jolla, California, where he at times found two octopuses under a single rock. His experiment similarly consisted of keeping two octopuses in a single tank. The solitary reputation of octopuses makes this a poor idea, and it is seldom done. Indeed, in some cases, one octopus would harass the other persistently, sometimes to death. However, if introduced at the same time, some pairs coexisted. These cases particularly interested Bateson.

Coexistence began with minor battles in which neither octopus was badly injured, a sort of testing phase. The larger octopus stole food from the smaller, and drove it out of shelter. After an interval, the smaller cautiously approached the larger—a dangerous move—but the larger then retreated. As Bateson saw it, this sequence established trust. First, the stronger octopus demonstrated strength. The weaker then showed its vulnerability by approaching regardless. Finally, and critically, the stronger then held back and refrained from harming the vulnerable octopus, as though showing "I can hurt you but I will not." From this point, the two octopuses could coexist without fighting, and sat in close proximity, sometimes touching.

Armed with these observations, in the final and most tense days of the Cuban missile crisis, Bateson then wrote a remarkable letter, seeking to bring to the attention of the Kennedy administration parallels between the international nuclear crisis and the behaviors of octopuses. The letter was to Bateson's colleague and mentor Warren McCulloch, who Bateson felt could direct the ideas to another colleague in the President's Science Advisory Committee and thereby reach policy makers within the Kennedy administration.

The Cuban Missile Crisis was resolved within a few days of the letter being written, so there was little time to act on it and no evidence McCulloch ever did. In the near aftermath, however, Bateson remarked that Kennedy had placed "trust" in Khrushchev's judgment, as the quarantine might have given Khrushchev a cause to be offended, but one that the Soviet ruler could decline to act on. That is, Bateson felt that Kennedy's quarantine of Cuba had provoked the Soviets in just the way one octopus could provoke another. The quarantine blocked only weaponry, and fell short of an air strike on the missile sites, or a blockade of Cuba, either of which would have been an act of war. The quarantine provided Khrushchev aggravation not conciliation. Would Khrushchev break the quarantine and land the missiles in Cuba anyway? But six Soviet ships containing weapons stopped short or reversed course before meeting the quarantining US forces. Khrushchev refrained. He subsequently agreed to remove the existing warheads from Cuba. An operational trust had been obtained allowing coexistence.

Bateson's observations were from captive octopuses interacting in pairs, and they described behaviors that remain rare or unheard of in octopuses, such as face-to-face mating, backing mantle-first toward a rival, and embracing one another after making peace. Out of concern for their welfare, captive octopuses are so seldom housed together that few independent observations exist to expand on Bateson's account. Where we find octopuses together in the wild, they are busily interacting with one another in complex ways. While some of these interactions escalate into battles and can be fatal, most are mediated by relational communication such as signals and low-intensity aggression that fall short of all-out hostility.

---

### Underwater, South of Sydney, Australia

TWO GLOOMY OCTOPUSES in adjacent dens at Octopolis both faced in the same direction, toward the outer edge of the site, into the void from where danger may approach. One was the octopus Quidnunc, a male. Possibly, if he could, he would monopolize all the females, and drive off all rival males. Next to him was a large female, just a half-arm's length away.

He reached toward her. She lifted an arm in response, perhaps in irritation at his approach. The Quidnunc pulled back. Then, adjusting his posture, he extended the third right arm, his mating arm. The second arm on that side lay over the extended third right. The mating arm turned pale, while his body pattern elsewhere was a more usual maroon splashed with cream. The other arms were dark above, lined below with orange and black just over the suckers. His eyes were dark, with black bars extended downward.

The curled tip of the third right arm unrolled toward the female, and extended. She tolerated this approach. The coupling was now out of view below the rim of her den and inside her gill slit, but we believe they began to mate. Still, the male's watchful eyes were elevated.

At the far edge of the habitat, hazy through a few meters of water, another octopus approached from off-site. Its silhouette stood darkly in contrast to the background of green water aglow with scattered daylight. The Quidnunc left off his mating, hastening to the far edge of the site. The mating had been brief: much shorter than uninterrupted matings. By allowing this interruption, it appeared the Quidnunc sired no eggs.

The approaching octopus fled back into the void. The Quidnunc stood tall and dark on spread arms at the edge of the site, leaning out toward the fleeing animal, as baleful as an approaching storm front, and swayed.

This was a distinctive pose, the same one displayed to the arriving females returning from foraging. He extended his web downward. His body color was dark. The mantle was elevated and rigid, sometimes only to the horizontal, sometimes straight up above the eyes. My colleagues and I informally call this "*Nosferatu,*" referencing the glowering specter of the 1922 German silent film.

Nosferatu poses may be more or less severe—less tall, less dark, mantle lower, arms less spread. They exhibit a continuum. The Nosferatu is associated with aggression; octopuses in this pose are likely to pursue another. Other octopuses are attentive to this, and may hesitate in their approach or redirect themselves. When one octopus matches the intensity of a Nosferatu pose by another, the interaction may escalate into a fight. At its extreme, the pose is large, high, dark, and intimidating. The Nosferatu is a signal to the other octopus, conveying a willingness to stand ground and not to back down. A signal, in this sense, is a behavior evolved to elicit a response in another. Signals emphasize features, such as vigor or size—"*I am tall; I am big*"—that carry their message.

There is a different signal for nonaggression, which conveys the opposite. The octopus presses flat to the ground. The arms and web are spread, but the arms curl into neat spirals. The mantle is low. The web and mantle display a high contrast pale and dark banding. This low display includes the low and spread body position of the deimatic pose, but differs in body color and usage.

OCTOPUSES COMING FROM OFF-SITE may approach in this low posture just described toward a stand-tall octopus. In another example, an octopus on the edge of the site crossed to the other side. The Quidnunc noticed the movement, and raised up to watch. He approached. The approach brought the Quidnunc close to a female in her den, interrupting her repose. She gathered a web full of debris and silt from her den and ejected it at the Quidnunc with a blast from her siphon. The Quidnunc now stood even taller, raising his eyes above the ensuing silt cloud to keep the crossing octopus in sight.

Noticing this continued attention, the crossing octopus displayed a more and more intense low display, wide and flat to the ground, with conspicuous pale banding across the neatly splayed web along each coiled arm. The Quidnunc was not appeased, however. Now displaying a strong Nosferatu, the Quidnunc steadily approached the octopus in low display. Both abandoned this posturing as the approach turned into a pursuit; the crossing octopus leapt off the substrate and jetted in pale guise around the periphery of the site, pursued by the dark and tempestuous Quidnunc.

SIGNALS WERE NOT the only behaviors of note in the three-part interaction just described. Maybe you noticed it—a female gathered up a bolus of silt and shells, and directed it through the water at a male. This is throwing, a form of tool use. Throwing at a conspecific is a social interaction—projectiles in this context are agonistic social tools. Nonhuman animals throwing objects at targets is relatively rare; species that throw projectiles at conspecifics as the targets are rarer still.

However, it is challenging to demonstrate targeting by nonverbal animals that do not say toward where they aim. My collaborator, Peter Godfrey-Smith, is a philosopher of science and an admirably careful thinker. He noted some of the challenges in claiming that an animal was targeting something. The idea of targeting is that the thrower *intends* the projectile to go somewhere particularly, whether or not the thrower achieves that aim. Since the targeting requires intention, an interior subjective state, it is difficult to demonstrate. However, there are clues.

To throw, octopuses gather material in their arms and hold it under the web. They then contract the mantle and forcefully eject water through the funnel positioned under the web, and at the same time release the debris, which is expelled under pressure. The octopus often rears up in doing so. The material is projected through the water up to several body lengths from the thrower.

Octopuses throw in three different contexts. They may throw away debris when den cleaning, throw away prey remains after eating, and throw material at other octopuses nearby. These latter throws in the direction of another octopus sometimes hit it. Throws occasionally are directed slightly to the side (a clue to targeting) and emerge between the right (or the left) first and second arms. At Octopolis in our videos, these slightly side-armed blasts are more likely to hit an apparent target.

Throws within range of another octopus often looked targeted, different from the throws after eating and while den cleaning. Octopuses threw shells, silt, and occasionally, algae. Shells were thrown many times when another octopus was nearby, but shells were an even more commonly thrown material after eating or when den cleaning. When another octopus was nearby, silt was more often thrown than it was in the other contexts. Throws near another octopus were more vigorous than den cleaning throws. All of these suggest that throwing is directed at other octopuses. We also found that these higher vigor throws were accompanied by dark color, which signals a willingness to stand ground.

Targeting has a role in interactions between octopuses. Females throw at other females, and these throws hit, while in our sample of one hundred throws only one throw by a male hit another octopus. As exemplified in the three-way interaction noted previously, females also may throw at males. The thrower may hit the target with flying shells or envelop him in an obscuring cloud of silt. Such hits occur within these bouts of ongoing signals, mild aggression, and jostling.

Throwing alters the target's behavior. A throw can bring the targeted octopus up short. A target may duck just before the throw is released, in a way that reduces or avoids the impact of the projectiles. Sometimes the target raises an arm in the direction of the thrower.

Octopuses may use puffs or jets of water from the siphon to displace small fish from bothering them. In other cases, they may reach out with one arm to swat away a fish. The thrown silt clouds step this up a bit, projecting flying shells or gritty silt through the water column. This can discourage the target octopus from approaching or disturbing the thrower. For example, a female may throw at a neighbor who is den cleaning and shoving debris from her den toward the thrower's den, or she may throw at the Quidnunc roaming the site as he travels too near. The offending female may then interrupt her cleaning, or the Quidnunc may redirect around the silt cloud. Throwing at conspecifics, as these octopuses apparently do, is rare in the animal world; to date, it has only been reported among social mammals (specifically, chimpanzees, capuchin monkeys, and dolphins).

As I described in both earlier incidents, Nosferatu and low displays almost always appeared in our video, with varying intensity, when two octopuses approached one another. Arriving females or males, ones that entered from off-site, most often responded with low displays to the Quidnunc's Nosferatu. These arriving (or returning) females, once close to the Quidnunc, alternatively may also show elements of the Nosferatu, as do females established in their dens—a sensible arrangement, as neither the female nor the Quidnunc is likely to strictly object to the presence of the other. Males confronting another male usually also engaged in unmatched displays: one in a Nosferatu, the other in a low display. Which male adopted which posture, however, depended on their history and circumstances.

The Quidnuncs, whether or not arriving from off-site, stood tall in varying intensities of Nosferatu in nearly every encounter. The Quidnunc can be busy, constantly challenging males that persisted in approaching the site. Sometimes one or more of these males occupied dens on the periphery, where the Quidnunc lost track of them, at least temporarily. Other times, a Quidnunc forcibly evicted a peripheral male from his den. These males that interacted with a Quidnunc showed the low display; but some of them, interacting with another non-Quidnunc male, would stand tall in a partial Nosferatu.

At Octlantis, where the rocky outcrops are broken into three patches at distances of several meters from each other, local Quidnuncs one year established on each patch, displaying Nosferatu to arriving octopuses. These local Quidnuncs kept to their particular patches, too busy to encounter the octopuses at other patches. Evicted octopuses, presumed male, repeatedly approached each patch in low display, to be met with the opposing Quidnunc in Nosferatu. As night fell on our videos and data collection that year, I hoped the evicted males were able to enter dens again somewhere on the periphery of the patches. It is dangerous out there in the dark.

In an interesting twist on expectations, some males in the peripheral dens were involved in few such encounters, and instead made the most of daytime mating attempts. These quiet males, with dens adjacent to females, went unnoticed by the Quidnuncs who, by contrast, were involved in few daytime mating attempts. What is all the activity for if the active, visible, threatening-looking males are not garnering mating attempts? We do not know yet, but it is worth noting that these are daylight records. It is possible that the daytime jockeying by males secures access to positions near females, and that the important mating occurs at night. As our cameras do not record in the dark, and lighting the nighttime scene would create predation risk, we do not yet know what happens between dusk and dawn.

............................................

THERE IS A SURPRISING AMOUNT of activity going on at Octopolis and Octlantis, these small assemblies of very active mollusks. Each octopus arrives at the same location, but there finds different circumstances, and the octopuses are flexibly adapted to meet these in complex ways. The Cold War probing of relationships, the signaling of intentions that reduce hostility, and the tolerance of coexistence are everywhere apparent in these aggregations.

Evolution has shaped in octopuses many of the foundations for complex relationships. There are even limited hints of empathy with octopuses. My students note that when they splay their hand on the outside aquarium glass, the octopus will come to cover their hand on the watery side, looking them in the eye. This type of moment was captured in the movie *Arrival*, when the linguist charged to open communications with the aliens places her hand on the barrier separating the human and otherworldly environment. On the other side, an alien splays its very octopuslike heptapod limb. Laurel and Thursday shared this greeting daily.

...................

WILD OCTOPUS AGGREGATIONS have long been known, and explained, by patchy and concentrated opportunities for shelter. Indeed, the examples in this book may all be regarded in that light. Yet octopus aggregations are still surprising, particularly in contrast to the prevailing reputation of octopuses as solitary creatures. It is the sophisticated behaviors by which octopuses cope with each other's proximity that fascinated Gregory Bateson, and fascinates me.

If you visit Octopolis in the Australian winter, there may be only one or two octopuses there. The population is seasonal and thrives in the summer. The octopuses live only a year or so, and each year the site is repopulated by those coming of age.

Over the years, the central artifact has sunk, a bit. New sediment is deposited all the time, and gradually, either the dense metal is submerging into the substrate, or the sediment is building up around it, perhaps stabilized by the shell bed itself. The original seed for the site may disappear altogether. New silt may cover the shells. Only the relentless work of the octopuses can build the shell bed faster than it is covered by silt or is scattered by winter storms.

...................

WE HAVE LEARNED OF two Larger Pacific Striped Octopus colonies in the Pacific waters off Central America in the last half century. Each colony persisted for a while, but then seemingly vanished again. Are these ephemeral aggregations, temporary gatherings?

........................................

JOCK MACLEAN, in the late 1950s, found giant Pacific octopuses in colonies off Vancouver Island, beds where he could harvest a thousand pounds of octopus in a morning, at one location. He felt that the really big octopuses did not live within the colonies—at that size, who needs a den? Instead, the true giants lurked below the assemblage, eating the smaller residents as they came and went from the site.

Over the years, I have heard of one or two locations off British Columbia where such aggregations occur. With the help of local divers, I arrived at such a bed, or at least, a location along a small drop-off where occupied dens were close enough together to find several in one dive. None of the octopuses were even close to one hundred pounds in size. The octopuses remained in their dens, and I did not see any of them interact with each other, as each den was just out of sight of the next. However, for the gift of a crab, one curious or hungry individual came out to swim with me for a moment. Clearly, when active, even the solitary giant Pacific octopus, the first species that I had begun to learn about, might encounter one another more commonly than their otherwise solitary nature had led us to believe. Surely they, too, were adept at managing these community relationships, just as in captivity where these giants seem to see themselves in relation to their neighbors, recognize their caretakers, and express their like and dislike of individuals.

Jock MacLean's beds, however, also seem lost to our modern times. At least, after inquiring about such locations with dive communities in British Columbia, the site I visited was a faint echo, at best, compared to his tales.

........................................

THIS BOOK INTRODUCES barely a score of octopuses out of hundreds of species, whose differences are worth examining. In Alaska, my students and I encountered a large octopus species no one had noticed before—this novel species is related to, but genetically distinct from, the commonly found giant Pacific octopus. It looks different, bearing a frill around its mantle that is lacking in the more familiar giant. The novel species appears at greater depths than does its relative, and may be more often found on silty slopes. Possibly, as the warming oceans drive the giant Pacific octopus away from shallow coastal habitats, these two species will encounter one another more in the colder Alaska deeps. Like this species, many octopuses have yet to be the focus of more than one or two scientific studies, so that our understanding of octopuses returns again and again to the few most familiar or most curious species.

We see the interdependence of social animals, the chimpanzees or lions that are drawn to live together in family units, parent with offspring and siblings with cousins. No such forces act on most octopuses. Typically, the parents expire as juveniles hatch, and planktonic paralarvae are dispersed on the waves. Yet the concentration of shelter in areas of plentiful food sometimes return the octopuses to each other's company. Here then is another model of animals in association—not in the way of cooperating families, but of independently competent individuals navigating the demands of living with neighbors.

These aggregations of octopuses, perched at the base of marine cliffs or in temporary shell beds, also hang at the edges of our knowledge, visited for short periods of collection or study, but otherwise half-legend. Rare and ephemeral, these locations tantalize us with the potential of their residents.

Alone, octopuses connect themselves in ways far beyond the predator and prey of energy transfer. Their ecology interweaves, woof and warp, in life's great variety, so that each thread pulls on others, wrinkling interests and diverting attention. The octopus in supposed solitude even so draws throngs. A bustling center arises around them of associates, cohabitants, and neighbors. These attract still others—those seeking shelter, the diggers and winnowers, and the schooling masses. Passersby appear, hungry or merely curious, the groupers, sharks, and their kind. As from another world, divers and scientists drop in.

During my last dive on the Great Barrier Reef, a charming day octopus spent her time with me as part of her entourage along the submerged slopes of Heron Island. My air was low; it was time to leave. She was still by then, papillate and spotted, tucked into a nook on a coral bommie. I began my slow leisurely slide toward the surface, the bright bubbles of my exhalations in their eagerness racing ahead of me as I mentally bid her farewell. From my rising vantage, the octopus was there but unseen, so perfect was her camouflage. She is with me still—each sucker questing even in repose, my every thought a colored, textured glissade across skin—so tightly are we bound in a world of beings like her, like me, in a community of fellows on our small blue-pearl planet Earth.

# ACKNOWLEDGMENTS

## Alaska Native Peoples, Lands and Indigenous Knowledge

I thank the *Exxon Valdez* Oil Spill Trustee Council for their work to designate community facilitators for all restoration work done with Alaska Native communities that, in turn, provided a process for me to learn from Alaska Native elders. All of my community visits from 1995–1998 were arranged and conducted through this process. I am also grateful to the University of Alaska Anchorage (UAA) and Alaska Pacific University (APU) for supporting my participation as a Faculty Fellow in their 2009 Difficult Dialogues workshop on Alaska Native ways of teaching and learning, and on working with cross-cultural differences.

Thank you to Mike Eleshansky of Chenega, who returned with me to the shores of Old Chenega. Jerry Totemoff kindly gave me permission to write about our searching for octopuses in Tatitlek. Thanks to Simeon Kvashnikoff of Port Graham for his contributions, and for verbal permission at the time to take notes and write about our conversation one evening in his living room in Port Graham; and to Apela Colorado for her wisdom and discussions of my references to Indigenous cultural materials in my work, and for permission to describe our conversation surrounding the story of the hunter and the octopus.

Thank you to Dee Pletnikoff (office manager), John Johnson (vice president of Cultural Resources), Tatianna Turner (cultural coordinator) and the Chugach Alaska Corporation for help reconnecting with those I worked with, or their families, in the Alaska Native communities. Thank you to John Johnson for discussing with me my account of

Simeon Kvashnikoff's tales in Port Graham, and for permission to retell Annie and Galushia Nelson's tales of giant octopuses near Cordova, which also appeared in Johnson's book *Eyak Legends of the Copper River Delta Alaska*. I am grateful to Cheryl Eleshansky for permission to write about her father, and my visit with him to Old Chenega and his account of the earthquake and its aftermath; and to Chuck Totemoff (chairman and president) of Chenega for permission to publish my portrayal of the 1964 earthquake and tsunami at Old Chenega.

Thanks to Alaska Native Language Center, University of Alaska Fairbanks, and to Professor Walkie Charles (director) and Leon Unruh (editor) for permission to use ANLC published content in my work, and for approval of my account of my language interview with Michael Krauss and Jeff Leer (Chapter 4). Thank you to the family of Robert Cogo (Robert and Chas Edwardson, and Skíl Jáadei [Linda Schrack]) for permission to retell the story of Raven woman and the octopus (Chapter 5).

Thank you to Michael Livingstone, Beth Leonard, Karli Tyance Hassell, and James Tempte, Apela Colorado, Juniper Scheel, and Peter Godfrey-Smith, each of whom cared enough to remind me where I had failed to heed the lessons I already knew regarding Alaska Native and other aboriginal peoples, and to encourage me to seek respectful permission to include their stories in writing about octopuses. Indigenous cultures own their stories. History did not begin with European or Russian explorations around the globe. Alaska Native people are with us today and continue to practice their cultures; they have not vanished. Remaining errors, omissions, and missteps are my own. It is my hope and intent to celebrate, respect, acknowledge, and credit the contribution of Alaska Native communities, and the aboriginal peoples and lands of Jervis Bay, Australia, and the Vezo people of Andavadoaka, Madagascar, to my own experiences understanding octopuses.

I have tried to relate the stories from other cultures as I find them, but I've told them in my own words rather than take verbatim quotations from sources. I hope to honor the storytelling tradition and story work that relates a given story to my own and each reader's experiences, with contextual references that are specific to octopuses, rather than treating traditional story translations as linear or Westernized plots. Therefore my retelling of these cultural stories should be uniquely meant for the intended readers of this book. However, the retelling necessarily blends my own style with that of the culture from which the story is drawn.

I relied on published translations or English-speaking narrators for all stories; errors of translation and Western adaptation no doubt are present, including my own divergence from source text. Sources are acknowledged throughout. I comment on the cultural role of these stories only in the words of the native Alaskans and the aboriginal people I worked with in the context of a given story. Inevitably, however, the cultural worldview of the source culture is obscured, misrepresented, or lost in translation to English. I see the stories and their meanings through my own light, and I relate those here in working with octopuses.

### Research Teams

Thank you to the wonderful students, whose keen interest and bubbling enthusiasm result in many great conversations about octopuses and the things they do. A special thank you to the many student aquarists, the fall Aquarium Biology classes, and graduate student coordinators of the Aquarium Lab at Alaska Pacific University for the assistance with animal care.

Thank you also to Captain N. Oppen of the research vessel *Tempest* for his professionalism, hard work, dedication to the long-term success of these surveys, and for his good humor in the field; and to fellow researchers T. L. S. Vincent, students of the annual summer octopus expeditions 2001–2016 at Alaska Pacific University, and to all my dive buddies for their underwater professionalism, companionship, and good cheer.

In Australia, I am indebted to Matt Lawrence, who discovered Octopolis and provides inspiration and logistical improvements to all our work in Jervis Bay, to Martin Hing and Kylie Brown, who discovered Octlantis, to Peter Godfrey-Smith, who first invited me to visit the site, and to Stefan Linquist and Stephanie Chancellor for all the conversations about octopuses and for their good company and contributions in the field.

In Mo'orea, I thank Jennifer Mather, Tatiana Leite, and Keely Langford for inviting me to join their research team, sharing different approaches to the same questions, and for being collaborators with me.

I thank Blue Ventures, Charlie Gough, Bris, and the Madagascar research team for my opportunity to work in Andavadoaka, for reviewing the chapter on Velondriake, and for permission to write about Bris's underwater search for the trout.

### Writing

Thank you to the Anchorage Title Wave Writers without Titles, who commented on early drafts of this book and provided a test audience: Celeste Borchardt, Lizzie Newell, Aileen Holthaus, Richard Herron, Les Tubman, Mary Edmunds. Other readers of early drafts also offered suggestions for improvements, including Juniper, Laurel, Edward and Griffin Scheel, and Peter Godfrey-Smith, as well as my editor at W. W. Norton, John Glusman.

Thank you to Tania Vincent, who participated in many years of the field work in Prince William Sound, and planned with me the first conception that would eventually lead to this book. I am also grateful to Sy Montgomery, who read the early work and pushed me toward publishing, including recommending me to my agent and urging me to get started on the project after my documentary with PBS and the BBC was released. I thank my agent Leslie Meredith, who encouraged me to revise my book proposal and who found interested publishers to consider my work.

Many people helped me chase down facts or consider how to present the science of octopuses. Thank you especially to those who answered questions or reviewed earlier drafts of chapters: Eric Chudler (neurons), Dominic Sivitilli (neurobiology), Terri Sheridan, Vanessa Delnavez, and Richard Smalldon at the Santa Barbara Museum of Natural History, Marla Daily at the Santa Cruz Island Foundation (size records of large octopuses), and Charlie Gough of Blue Ventures (Madagascar). The book is better for their help; the remaining errors are my own. I am grateful to the cephalopod research community for being fascinating, welcoming, and energizing professional colleagues; it has been a great pleasure to associate with this community over my career.

I am especially indebted to my lifelong and closest friends RLB and GSIII, and to my family. Throughout my career, they have been my closest intellectual companions outside of the academic world. They have kept me at home (to the extent I am) in the nonspecialist world, and been my companions exploring beauty's ultra-fringe.

# NOTES

## INTRODUCTION: THE INNER LIVES OF OCTOPUSES

3  *a seismic shock of nearly 7.0 magnitude:* Project Cannikin reported a seismic disturbance "body wave magnitude of 6.8." The blast temporarily lifted the surface above the
test site by twenty-five feet, and permanently raised the nearby beach and ocean floor
by four to six feet.

The Cannikin test and other Alaska nuclear tests and plans were controversial.
These were nuclear devices larger than those detonated at Hiroshima (about fifteen kilotons) and Nagasaki (about twenty kilotons). The blasts beneath Amchitka,
located along an active fault zone, might trigger tsunamis or large earthquakes or leak
radiation. These concerns led in 1969 to the formation of the Don't Make a Wave
Committee antinuclear protest organization, which within two years evolved into
Greenpeace.

The weapons test and its aftermath remain controversial. The 2011 report noted
here includes the results of our Amchitka kelp sampling efforts—radiation levels at
Amchitka were not above what the US Department of Energy considers a very low
and acceptable increase in risk to human health outcomes.

Project Cannikin, Amchitka Island, November 6, 1971. Atomic Tests Channel:
https://www.youtube.com/watch?v=1JJEPBLL4E8.

Miller, P., and Buske, N. "Nuclear Flashback: Report of a Greenpeace Scientific
Expedition to Amchitka Island, Alaska—Site of the Largest Underground Nuclear
Test in U.S. History." 1996. http://www.fredsakademiet.dk/ordbog/uord/nuclear_
flashback.pdf.

Hunter, Robert. *The Greenpeace to Amchitka: An Environmental Odyssey.* Arsenal
Pulp Press. 2004.

Legacy Management. Amchitka Island, Alaska, Biological Monitoring Report
2011 Sampling Results (Ed. USDo Energy). 2013.

3  *questions that had not, by then, yet received definitive study:* For a brief review of the

question of hearing by octopuses, see Hanlon, R. T., and Messenger, J. B. *Cephalopod Behaviour*, 2nd ed. (New York: Cambridge University Press. 2018).

4 ***cannibals, with limited means to recognize their fellow beings:*** A pretty good summary of this perspective of octopuses appeared only twelve years before this writing in: Mather, Jennifer A., Anderson, Roland C., and Wood, James B. *Octopus: The Ocean's Intelligent Invertebrate* (Timber Press, 2010). (See page 134.)

A similar summary again appears in: Dölen, G. "Mind Reading Emerged at Least Twice in the Course of Evolution." In Linden, G. *Think Tank: Forty Neuroscientists Explore the Biological Roots of Human Experience* (New Haven, CT: Yale University Press, 2018), 194–200. (See page 198.)

## CHAPTER 1: STARTING OUT IN ALASKA

13 ***mud-flat clam beds above the high tides:*** Described by Henry Makarka in *The Day That Cries Forever: Stories of the Destruction of Chenega during the 1964 Alaska Earthquake,* collected and edited by John Smelcer (Anchorage, AK: Chenega Future Inc., 2006).

14 ***Eyak village was annexed to the town in 1900:*** The Native Village of Eyak maintains a website, https://www.eyak-nsn.gov/. A brief account of the impact of the death of Marie Smith Jones can be found at, https://www.alaskanativelanguages.org/eyak. The daXunhyuu (Eyak people) language revitalization web page is at https://www.alaskanativelanguages.org/eyak.

## CHAPTER 2: DANGEROUS GIANTS

16 ***according to reports from Washington State:*** High (1976) noted "most" weigh less than 70 pounds. Hartwick (1984), reviewed their biology and reported sexual maturation for females at 10 kg (22 pounds) weight, for males at 15 kg (33 pounds).

High, W. L. "The Giant Pacific Octopus." *Marine Fisheries Review* 38, no. 9 (1976): 17–22.

Hartwick, E. B. *Octopus dofleini.* In "Cephalopod Life Cycles, Vol. I. Species Accounts." (Ed. P. R. Boyle) (London: Academic Press, 1983), 277–91.

17 ***too close to a big octopus that grabbed him by one leg:*** Kaniut, L. *Cheating Death: Amazing Survival Stories from Alaska* (Fairbanks, AK: Epicenter Press, 1994).

17 ***they were all bad, and they ate people*** and *for the water was too thick with its slime:* "Giant Animals" (Tale 15A) and "The Giant Devilfish" (Tale 15B). In Birket-Smith,

Kaj, and de Laguna, Frederica. *Eyak Indians of the Copper River Delta, Alaska* (Copenhagen: Levin and Munksgaard, 1938).

18 ***an 1897 account from the Smithsonian Institution:*** Verrill, A. E. "The Florida Sea-Monster." *American Naturalist* 31 (1897): 304–7, reproduced in Appendix C: A. E. Verrill's account of the Florida Sea-Monster. In Ellis, R. *Monsters of the Sea* (New York: Alfred A. Knopf, 1994).

18 ***Indigenous cultures:*** The term *Indigenous* is capitalized when referring to the original people of a land, but not, for example when referring to indigenous flora or fauna: https://www.sapiens.org/language/capitalize-indigenous/.

18 ***how would I get into the water with them and work safely?:*** As I write this decades later, a recent paper reports the largest recorded individual of *Enteroctopus dofleini*, the giant Pacific octopus, at 198.2 kg (437 pounds). See details in the chapter about this questionable record. The paper author (McClain) is a deep-sea biologist who became interested in the evolution of body size in the deep sea as a function of food availability.

McClain, C. R., Balk, M. A., Benfield, M. C., Branch, T. A., Chen, C., Cosgrove, J., Dove, A. D., Gaskins, L. C., Helm, R. R., and Hochberg, F. G. "Sizing Ocean Giants: Patterns of Intraspecific Size Variation in Marine Megafauna." *PeerJ* 3 (2015): e715.

19 ***current and more accurate maximum estimates of forty-two-feet length and 500 kg (1,100 pounds) are impressive enough:***

Length: O'Shea, Steve, and Bolstad, Kat. "Giant Squid and Colossal Squid Fact Sheet." *TONMO: The Octopus Online News Magazine.* 2019: https://tonmo.com/articles/giant-squid-and-colossal-squid-fact-sheet.18/.

Weight: Jereb, P., Roper, C. F. E. *Cephalopods of the World. An Annotated and Illustrated Catalogue of Cephalopod Species Known to Date. Volume 2. Myopsid and Oegopsid Squids.* FAO Species Catalogue for Fishery Purposes 2, no. 4 (2010): 605.

The colossal squid (*Mesonychoteuthis hamiltoni*, Robson, 1925), most commonly found in the deep Southern Ocean, is as heavy or possibly heavier than *Architeuthis*, reaching at least 495 kg (1,089 pounds).

Rosa, R., Lopes, V. M., Guerreiro, M., Bolstad, K., and Xavier, J. C. "Biology and Ecology of the World's Largest Invertebrate, the Colossal Squid (*Mesonychoteuthis hamiltoni*): A Short Review." *Polar Biology*, 40, no. 9 (2017): 1871–83.

19 ***to follow them to the giant squid, their quarry and his:*** Clyde Roper's critter cam on sperm whales would not succeed in capturing footage of live giant squid. The first such

footage was obtained from a baited deep-sea mid-water camera station ten years later by Japanese researchers.

Kubodera, T., and Mori, K. "First-Ever Observations of a Live Giant Squid in the Wild." *Proceedings Royal Society of London* B 272, no. 1581 (2005): 2583–86.

19 *the largest animal ever to live:* The largest blue whale reaches 32 m. Early size estimates of the now-famous megalodon were as large. However, revised estimates of the largest megalodon, improved with careful attention to shark proportions, are now at 15 m.

Sears, R., and Perrin, W. F. "Blue Whale: *Balaenoptera musculus.*" *Encyclopedia of Marine Mammals*, 2nd ed. (London: Academic Press, 2009), 120–24.

Shimada, K. "The Size of the Megatooth Shark, *Otodus megalodon* (Lamniformes: Otodontidae), Revisited." *Historical Biology* (2019): 1–8.

21 *retraction of his identification of the carcass as a Colossal Octopus:* Verrill, A. E. "The Florida Sea-Monster." *American Naturalist* 31 (1897): 304–7, as reproduced in Appendix C: A. E. Verrill's account of the Florida Sea-Monster in: Ellis, R. *Monsters of the Sea* (New York: Alfred A. Knopf, 1994).

21 *an account of their explorations:* Wood, F. G. "An Octopus Trilogy. Part 1: Stupefying Colossus of the Deep." *Natural History* (March 1971): 15–24.

Gennaro, J. F. Jr. "An Octopus Trilogy. Part 2: The Creature Revealed." *Natural History* (March 1971): 24, 84.

For other related accounts, including a brief note about Wood and Gennaro's examination of the tissue, see: Mangiacopra, G. S. "The Great Ones: A Fragmented History of the Giant and the Colossal Octopus." *Of Sea and Shore* 19, no. 1 (Summer 1976): 93–96.

21 *the tissue most resembled that of an octopus:* Mackal, R. P. "Biochemical Analyses of Preserved Octopus Giganteus Tissue." *Cryptozoology* 5 (1986): 55–62.

21 *dubbed the Bermuda Blob:* Greenwell, J. R. "The Bermuda Blob." *BBC Wildlife* (August 1993): 33.

21 *no evidence to support the existence of* Octopus giganteus: Pierce, S. K., Smith, G. N., Maugel, T. K., Clark, E. "On the Giant Octopus (*Octopus giganteus*) and the Bermuda Blob: Homage to A. E. Verrill." *Biological Bulletin* 188 (1995): 219–30.

21 *brief news article appeared in* Science . . . *moved to write a rebuttal:* Anonymous. "One Sea Monster Down." *Science* 268 (1995): 207–9.

Mangiacopra, G. S., Smith, D. G., Avery, D. F., Raynal, M., Ellis, R., Greenwell,

J. R., Heuvelmans, B. "An Open Forum on the *Biological Bulletin*'s Article of the *Octopus giganteus* Tissue Analysis." *Of Sea and Shore* 19, no. 1 (1996): 45–50.

22 **the remains of sperm and fin whales:** Carr, S., Marshall, H., Johnstone, K., Pynn, L., and Stenson, G. "How to Tell a Sea Monster: Molecular Discrimination of Large Marine Animals of the North Atlantic." *The Biological Bulletin* 202, no. 1 (2002): 1–5.

Pierce, S. K., Massey, S. E., Curtis, N. E., Smith Jr., G. N., Olavarria, C., and Maugel, T. K. "Microscopic, Biochemical, and Molecular Characteristics of the Chilean Blob and a Comparison with Remains of Other Sea Monsters: Nothing but Whales." *The Biological Bulletin* 206 (2004): 125–33.

22 **half of that protein is collagen:** Lockyer, C. H., McConnell, L. C., and Waters, T. D. "The Biochemical Composition of Fin Whale Blubber." *Canadian Journal of Zoology* 62, 1 no. 2 (1984): 2553–62.

22 **even after the oil in the blubber decays:** A male sperm whale may reach 55 tons. One estimate of the amount of a sperm whale that is blubber appears in *Moby Dick* (Chapter 68: "The Blanket"): "Reckoning ten barrels to the ton, you have ten tons for the net weight of only three quarters of the stuff of the whale's skin." Thus there would be 13.3 tons of blubber to yield this 10 tons of oil. Thirty percent of the blubber is estimated to be protein and half of that protein is collagen. From 13 tons of blubber, a 55-ton sperm whale would then leave behind 2 tons of pure collagen protein alone.

Melville's science in *Moby Dick* was based on his own experience as a whaler and on the best scientific sources available in his day (Olsen-Smith 2010, Zimmerman 2018); however, he also exaggerated for dramatic effect. Still, the numbers that Melville used in his ballpark estimate were accurate: Melville's yield from a "very large" whale provides an estimate that the blubber comprises 24 percent of the total weight of the whale (assuming a 55-ton sperm whale), well within modern estimates that blubber comprises 20–30 percent of the total weight, and for some species as much as 50 percent (Reynolds and Rommel 1999, Iverson 2009).

Iverson, S. J. "Blubber." *Encyclopedia of Marine Mammals*. (London: Academic Press, 2009), 115–20.

Olsen-Smith, S. "Melville's Copy of Thomas Beale's 'The Natural History of the Sperm Whale and the Composition of Moby-Dick.'" *Harvard Library Bulletin* 21, no. 3 (2010): 1–77.

Reynolds, J. E., Rommel S. A. *Biology of Marine Mammals*. Smithsonian Institution, Washington. 1999.

Zimmerman, C. "'Therefore His Shipmates Called Him Mad': The Science of Moby-Dick." *Medium*, January 9, 2018: https://carlzimmer.com/therefore-his -shipmates-called-him-mad-the-science-of-moby-dick-2/.

22 **the stench of rancid whale oil:** I am grateful to whale biologists Eva Saulitus (Homer, AK), Rod Palm (Tofino, BC), Christina Lockyer (Denmark), and Peter Arnold (Australia) for discussing with me in the 1990s their own encounters with decaying whales at sea that confirm this description, despite that dead whales can also, and perhaps more famously, sink intact to the ocean floor.

22 **The spermaceti organ of a sperm whale:** Sperm whale foreheads contain two large oil-filled compartments: the spermaceti organ and the junk sac, or junk. Together they constitute up to one-quarter of body mass and extend one-third of the total length of the whale. Both are recognized as important in echolocation. Spermaceti is a waxy substance that was an important fuel for candles, which burned brighter, longer, and cleaner than other materials. The junk may also protect the jaw and skull, permitting ramming combat between males competing for mates, and allowing these whales to survive the ramming of whaling and research ships such as the *Essex* (1820), the fictional *Pequod* (in *Moby Dick*, published 1851), and possibly the research vessel *Kahana* (in 2007), among many others.

Panagiotopoulou, Olga, et al. "Architecture of the Sperm Whale Forehead Facilitates Ramming Combat." *PeerJ* 4 (2016): e1895.

Fulling, Gregory L., et al. "Sperm Whale (*Physeter macrocephalus*) Collision with a Research Vessel: Accidental Collision or Deliberate Ramming?" *Aquatic Mammals* 43, no. 4 (2017): 421.

25 *from James Cosgrove, the chief of Natural History Collections at the Royal British Columbia Museum:* Jim's accounts of large octopuses can be found in his later book, Cosgrove, J. A., and McDaniel, N. *Super Suckers: The Giant Pacific Octopus and Other Cephalopods of the Pacific Coast.* (Madeira Park, BC: Harbor Publishing, 2009).

25 *At the Santa Barbara Museum of Natural History:* I was able to examine these two photographs with the help of Eric Hochberg, the curator of invertebrates, and Vanessa Delnavaz, the invertebrate zoology collection manager, of the Santa Barbara Museum of Natural History. Notes on the back of the image of octopus and fisherman read: "1945 Skipper Babe Castagnola catches large octopus at Santa Cruz Island weighing 402 lbs off the coast of Santa Barbara Calif Photo 2." The annotation on back of the other photograph reads "March 23 1912 Catalina Island. Courtesy of the Catalina

Island Museum, Avalon CA." For more octopus encounters from the Catalina Islands, see https://www.islapedia.com/index.php?title=Octopi. The lack of a specific date on the back of the 1945 image might mean that it was acquired by the museum much later than the catch. I was unable to find an account of the catch and weighing of this octopus contemporaneous with those events.

25   *In one account of Jock's tales:* Newman, M. *Life in a Fishbowl: Confessions of an Aquarium Director.* (Vancouver, Toronto: Douglas & McIntyre 1994.) Portions of this book can be accessed online at: https://archive.org/details/lifeinfishbowlco0000newm/mode/2up?q=octopus.

26   *in another version:* High, W. L. "The Giant Pacific Octopus." *Marine Fisheries Review* 38 (September 1976): 17–22 (MFR Paper 1200).

## CHAPTER 3: LOST HOMES

35   *I had not heard about the earthquake from someone who had been through it:* Until 2006, published accounts of the 1964 earthquake included first-person reports from Anchorage, Valdez, and several other Southcentral Alaskan communities, but no first-person accounts from the village of Chenega. At the time I wrote the first draft of this account (1996) imagining the devastation of Old Chenega; I based it on Mike Eleshansky's account told to me at the kitchen table that night, and later revised some details using firsthand descriptions and video of the 2004 Indian Ocean tsunamis that hit Indonesia and Thailand.

Those same 2004 tsunamis prompted an effort to collect firsthand stories of the 1964 earthquake from Old Chenega. These were published in *The Day That Cries Forever* (see the first endnote for Chapter 1). My imagined account of that day in Chenega has also been informed by firsthand descriptions published in that source, especially the recollections of Nick Kompkoff Jr.

## CHAPTER 4: OUR COUSIN OCTOPUSES

44   *the Oregon rock crab* (**Glebocarcinus oregonensis**)**:** The more familiar species name *Cancer oregonensis* is currently (and correctly) *Glebocarcinus oregonensis*. A revised classification (Schweitzer and Feldmann 2000) of the cancer crabs, in part based on discovery of new fossils, is used more among biologists now than when initially published. What once were subgenera have been elevated to the level of genus. *Glebocarcinus* genus is characterized by a carapace length of about three-quarters its width, often

with granular regions—a description that matches this Oregon rock crab. *Glebocarcinus* crabs probably evolved in the North Pacific, and now occur only around the North Pacific Rim.

> Schweitzer, C. F., and Feldmann, R. M. "Re-evaluation of the Cancridae Latereille, 1802 (Decapoda: Brachyura) Including Three New Genera and Three New Species." *Contributions to Zoology* 69, no. 4 (2000): 223–50.

45 ***paralarvae shift their behavior from swimming to clinging:*** Dan, S., Shibasaki, S., Takasugi, A., Takeshima, S., Yamazaki, H., Ito, A., and Hamasaki, K. "Changes in Behavioural Patterns from Swimming to Clinging, Shelter Utilization and Prey Preference of East Asian Common Octopus *Octopus sinensis* during the Settlement Process under Laboratory Conditions." *Journal of Experimental Marine Biology and Ecology* 539 (2021): 151537.

46 ***the technical taxonomic key:*** Kozloff, E. *Marine Invertebrates of the Pacific Northwest* (Seattle: University of Washington Press, 1987).

48 ***Michael was an expert in the Eyak language:*** Michael Krauss died at age eighty-five in 2019: https://abcnews.go.com/US/wireStory/michael-krauss-alaska-linguistics-expert-dead-84-65022379.

48 ***the word for octopus is* tse-le:x-guh:** This spelling was given by Michael Krauss, and matches his usage in:

> Krauss, M. E. (ed). *In Honor of Eyak: The Art of Anna Nelson Harry* (Fairbanks: Alaska Native Language Center, University of Alaska Fairbanks, 1982).

> The dAXunhyuu Learner's Dictionary uses the spelling *tsaaleeXquh* and provides a recording of the word's correct pronunciation in Eyak: https://eyakpeople.com/dictionary.

48 ***The Eyaks on the Copper River Delta (from Cordova eastward) are at the base of the thumb:*** For an excellent online map of Indigenous languages throughout the Americas (and in Alaska) and in Australia, see Native Land Digital: https://native-land.ca/.

> The Alaska portion of this map depends in part on a map originally drawn by Michael Kraus.

> Krauss, Michael, Holton, Gary, Kerr, Jim, and West, Colin T. "Indigenous Peoples and Languages of Alaska" (Fairbanks and Anchorage: Alaska Native Language Center and UAA Institute of Social and Economic Research, 2011).

49 ***three words for octopus:* amq̂u, aaqanax̂,** *and* **ilgaaq̂ux̂:** The Aleutian Pribolof Islands Association's Qaqamiiĝux̂ Film Series video titled "How to Cook Octopus"

depicts how these words are pronounced and ways to prepare octopus: https://www .apiai.org/community-services/traditional-foods-program/videos/.

50  *the Inuit-Yupik-Unangax̂ language family:* This is a distinct language family of New World Arctic and near-Arctic coastal peoples reaching from Greenland to the Aleutian Islands chain, and unrelated to other language families of North America. It has also been referred to as the Eskimo-Aleut language family.

## CHAPTER 5: OCTOPUSES OVERRUN

55  *a fisherman at Bexhill hauled up the first octopus:* At least two plagues of octopuses struck the southern coasts of England, in 1899–1900 and in 1950. Some details as written here were taken from an account of the 1950 occurrence, and fictionalized back to 1899.

Garstang, W. "The Plague of Octopus on the South Coast and Its Effects on the Crab and Lobster Fisheries." *Journal of the Marine Biological Association of the United Kingdom* 6 (1900): 260–73.

Rees, W. J., and Lumby, J. R. "The Abundance of Octopus in the English Channel." *Journal of the Marine Biological Association of the United Kingdom* 33 (1954): 515–36.

For a description of the North Atlantic oscillation and its biological effects, see: Mann, K. H., and Lazier, J. R. N. "Chapter 9: Variability in Ocean Circulation: Its Biological Consequences." In *Dynamics of Marine Ecosystems: Biological-Physical Interactions in the Ocean*, 2nd Edition (Malden, MA: Blackwell Sciences, 1996), 282–316.

57  *legends of the Haida and Tlingit peoples:* Tales with the elements depicted in the story about Raven woman at Haida Village have been retold several times in print. Possibly the first English language appearance was as told to John Swanton (1909) by Katishan, who was chief of the Kasq!ague'dî of Wrangell, Alaska. Katishan was "considered the best speaker at feasts" and "to have a better knowledge of the myths than anyone else at Wrangell."

As related here, I used details both from Katishan and from Cogo (1979). Cogo notes that he "heard these stories told a long time ago." Subsequent published retellings are traceable to one or the other of these accounts.

The two versions share many elements. In each, the chief's daughter marries a devilfish and her children return to their grandfather's village. The villagers offend the devilfish and the devilfish determine to exact their revenge, whereupon both the chief's daughter and a shaman warn the people of the coming attack. In each, the devilfish attack but are ultimately defeated, and peace is later restored. Differences include how the chief's daughter meets her devilfish husband, whether the chief or children in the village offend the octopuses, and whether the villagers or the octopuses start the killing.

Eastman, C. M., and Edwards, E. A. *Gyaehlingaay: Traditions, Tales, and Images of the Kaigani Haida: Traditional Stories Told by Lillian Pettviel and Other Haida Elders* (Seattle: Burke Museum Publications, University of Washington Press, 1991).

Cogo, Robert. 1979. *Haida Story Telling Time.* Susan Horton, Ed. Ketchikan Indian Corporation.

NB: This octopus story appears on *P.O.W. Report* as "The Great Octopus of North Pass": https://www.powreport.com/2019/11/the-great-octopus-of-north-pass.html. An English-language narrative of portions can be heard on CBC Radio, "Legends of the Old Massett Haida": https://www.cbc.ca/player/play/2335401544 (from 29:30 to 37:05).

Swanton, J. R. *Tlingit Myths and Texts Recorded by John R. Swanton.* Smithsonian Institution, Bureau of American Ethnology Bulletin 39 (Washington, DC: US Government Printing Office, 1909), 130–32. Reprinted by Native American Book Publishers, Brighton, MI, 1990.

59 **They told their octopus mother and father what had happened:** Octopus parents provide no care to the young after the female tends the eggs until hatching. Thus, this scene of the parents with their offended young is not biologically realistic. However, it is truer to the stories. In the Tlingit version, the daughter returns with her octopus husband and children to dine with her father the Chief. In the Haida version, a baby octopus comes to the village, struggles to escape children playing with it, and on hearing this the octopus people become angry over what was done to their child.

62 **The pelagic octopod Argonauta argo:** Details about *Argonauta argo* and the puzzle of *Hectocotylus* were from:

Okutani, Takashi, and Kawaguchi, T. "A Mass Occurrence of *Argonauta argo* (Cephalopoda: Octopoda) along the Coast of Shimane Prefecture, Western Japan Sea." *Venus* 41 (1983): 281–90.

Finn, J. K., and Norman, M. D. "The Argonaut Shell: Gas-Mediated Buoyancy Control in a Pelagic Octopus." *Proceedings of the Royal Society B: Biological Sciences* 277, no 1696 (2010): 2967–71.

Mann, Thaddeus. "Section 3.2 Cephalopods." In *Spermatophores: Development,*

*Structure, Biochemical Attributes and Role in the Transfer of Spermatozoa.* Vol. 15 (Berlin: Springer Science+Business Media, 2012).

Kölliker, A. "II. Some Observations upon the Structure of Two New Species of Hectocotyle, Parasitic upon *Tremoctopus violaceus*, D. Ch., and *Argonauta Argo*, Linn.; with an Exposition of the Hypothesis That These Hectocotylæ Are the Males of the Cephalopoda upon Which They Are Found." *Transactions of the Linnean Society of London* 1 (1846): 9–21.

Baron Cuvier, Georges. "Memoire sur un ver parasite d'un nouveau genre (*Hectocotylus octopodis*)" (Paris: Annales des Sciences Naturelles, Crochard Collection, 1829).

63 *species whose young hatch into the plankton:* The estimate of over a hundred thousand eggs from a single female was reported in Conrath and Conners (2014). Note that species whose young are benthic rather than planktonic lay many fewer eggs (see Sauer et al. 2019).

Conrath, C. L., and Conners, M. E. "Aspects of the Reproductive Biology of the North Pacific Giant Octopus (*Enteroctopus dofleini*) in the Gulf of Alaska." *Fishery Bulletin* 112, no. 4 (2014): 253–60.

Sauer, W. H. H., et al. "World Octopus Fisheries." *Reviews in Fisheries Science & Aquaculture* (2019): 1–151.

## CHAPTER 6: GLOBAL OCTOPUSES

68 *because a major research initiative at the Science Center involved plankton blooms:* Cooney, R. T., Allen, J. R., Bishop, M. A., Eslinger, D. L., Kline, T., Norcross, B. L., McRoy, C. P., Milton, J., Olsen, J., Patrick, V., Paul, A. J., Salmon, D., Scheel, D., Thomas, G. L., Vaughan, S. L., and Willette, T. M. "Ecosystem Control of Pink Salmon (*Oncorhynchus gorbuscha*) and Pacific Herring (*Clupea pallasi*) Populations in Prince William Sound, Alaska." *Fisheries Oceanography* 10, suppl. 1 (2001).

68 *much of the productivity of the ocean surface waters:* Growth of algae on the ocean floor is also significant. Corals, with symbiotic algae in their tissues, and large kelps are among the most productive ocean habitats. In the deep sea, chemosynthesis in hot vent communities also contribute to ocean productivity.

69 *They hunt the crab zoea and megalops:* When octopuses hatch from the egg, they are small but of adult form, except that the proportions are different than that of an adult. Octopuses have paralarvae—as these bear resemblance to the adult, they are not true larvae.

Crustaceans, like insects, have true larvae that undergo metamorphosis: the larval body form does not look like the adult. For crabs, the *nauplius* first stage consists of a head and tail; the *zoea* larval second stage swims with the thoracic (chest) appendages rather than a cephalic (head) appendage; the *megalops* third stage bears a long abdominal section on which abdominal appendages appear, and the eyes are large. Proportions shift somewhat as this stage continues and (for benthic species) juveniles settle to the sea bottom.

Both the crustaceans and the more familiar insects are arthropods, having in common an exoskeleton (cuticle) of chiton, jointed limbs, and a segmented body. Particularly among the crustaceans, the cuticle is further hardened by the addition of calcium carbonate from the water. Arthropods must shed and replace the rigid exoskeleton nearly continuously as they grow to full size.

Octopuses, however, are mollusks. They have neither exoskeletons nor the other arthropod traits of segmented body and jointed limbs.

70 ***around Norway and Sweden in the North Sea:*** Schickele, A., Francour, P., and Raybaud, V. "European Cephalopods Distribution under Climate-Change Scenarios." *Scientific Reports* 11, no. 1 (2021): 1–12. Note, this study also models the distribution of the common octopus expanding into the nearly enclosed Baltic Sea, which has salinity too low for cephalopods, but the models did not take salinity into account for octopuses (only for squid and cuttlefish).

70 ***ten miles across the mouth of Valdez Arm and ten miles up Columbia Bay:*** A detailed relief map of Prince William Sound can be found from Shaded Relief at http://www.shadedrelief.com/pws/. The map illustrates the retreating terminus of the Columbia Glacier, 1978 to 2019. Satellite imagery can be seen courtesy of NASA at: https://earthobservatory.nasa.gov/world-of-change/ColumbiaGlacier.

71 ***frozen crystals that had formed thousands of years ago:*** The oldest glacier ice recovered and dated from Alaska was thirty thousand years old; the oldest glacier ice in the Antarctic may be close to one million years old. The USGS provides answers to frequently asked questions about glaciers at https://www.usgs.gov/faqs/how-old-glacier-ice.

72 ***the correlation of cold salty surface waters with populations of octopuses and squid around the world:*** I summarized some of the literature on this correlation in:

Scheel, D. "Sea-Surface Temperature Used to Predict the Relative Density of Giant Pacific Octopuses (*Enteroctopus dofleini*) in Intertidal Habitats of Prince William Sound, Alaska." *Marine and Freshwater Research* 66 (2015): 866–76.

72 ***the warmest in history:*** Cheng, Lijing, et al. "Another Record: Ocean Warming Con-

tinues through 2021 despite La Niña Conditions." *Advances in Atmospheric Sciences* (2022): 1–13.

73 ***survival declined by 70 percent:*** Survival of *Enteroctopus megalocyathus* embryos, a southern hemisphere relative of the Alaskan giant Pacific octopus, drops 15 percent when the optimal temperature range is exceeded by even 1 degree. *Octopus maya* embryo survival dropped by 70 percent during current-day bottom temperatures around 30°C. With a further increase of just 1°C, no embryos survived.

Sauer, W. H. H., et al. "World Octopus Fisheries." *Reviews in Fisheries Science & Aquaculture* (2019): 1–151.

## CHAPTER 7: OCTOPUSES SEIZED

77 ***their community's traditional fishing grounds along Madagascar's west coast:*** Astuti, Rita. " 'The Vezo Are Not a Kind of People': Identity, Difference, and 'Ethnicity' among a Fishing People of Western Madagascar." *American Ethnologist* 22, no. 3 (1995): 464–82.

Marikandia, M. "The Vezo of the Fiherena Coast, Southwest Madagascar: Yesterday and Today." *Ethnohistory* 48, nos. 1–2 (2001): 157–70.

78 ***only 20 to 30 percent coral cover typically, and low fish abundance and biomass:*** Nadon, M. O., Griffiths, D., Doherty, E., and Harris, A. "The Status of Coral Reefs in the Remote Region of Andavadoaka, Southwest Madagascar." *Western Indian Ocean Journal of Marine Science* 6, no. 2 (2008).

Gilchrist, H., Rocliffe, S., Anderson, L. G., and Gough, C. L. "Reef Fish Biomass Recovery within Community-Managed No Take Zones." *Ocean & Coastal Management* 192 (2020): 105210.

79 ***export companies brought accessible commercial markets:*** Harris, A. " 'To Live with the Sea' Development of the Velondriake Community-Managed Protected Area Network, Southwest Madagascar." *Madagascar Conservation & Development* 2, no. 1 (2007).

79 ***sardines near the village, too, are seen less frequently now than in the past:*** Gough, C., Thomas, T., Humber, F., Harris, A., Cripps, G., and Peabody, S. "Vezo Fishing: An Introduction to the Methods Used by Fishers in Andavadoaka Southwest Madagascar." *Blue Ventures Conservation Report*. Blue Ventures, UK (2009). Available at: https://blueventures.org/publications/vezo-fishing-introduction-methods-used-fishers-andavadoaka-southwest-madagascar/.

79 ***mass distribution in Africa of fine-mesh bed nets as protection from malaria-***

*carrying mosquitos:* Jones, B. L., and Unsworth, R. K. "The Perverse Fisheries Consequences of Mosquito Net Malaria Prophylaxis in East Africa." *Ambio* (2019) 1–11.

80 ***higher than it had been without management:*** Oliver, T. A., Oleson, K. L., Ratsim-
bazafy, H., Raberinary, D., Benbow, S., and Harris, A. "Positive Catch & Economic
Benefits of Periodic Octopus Fishery Closures: Do Effective, Narrowly Targeted
Actions 'Catalyze' Broader Management?" *PLOS One* 10, no. 6 (2015): e0129075.

81 ***harvesters who depend on the sea:*** A review of world octopus fisheries can be found
in: Sauer, W. H. H., et al. "World Octopus Fisheries." *Reviews in Fisheries Science &
Aquaculture* (2019): 1–151.

81 ***octopus was observed enveloping a lionfish in its arms and web and subduing it:***
Crocetta, F., Shokouros-Oskarsson, M., Doumpas, N., Giovos, I., Kalogirou, S., Lan-
geneck, J., Tanduo, V., Tiralongo, F., Virgili, R., and Kleitou, P. "Protect the Natives to
Combat the Aliens: Could *Octopus vulgaris* Cuvier, 1797 Be a Natural Agent for the
Control of the Lionfish Invasion in the Mediterranean Sea?" *Journal of Marine Science
and Engineering* 9, no. 3 (2021): 308.

82 ***challenging in several ways:*** Ideas here are drawn from Franks, B., Jacquet, J., Godfrey-
Smith, P., and Sánchez-Suáre, W. "The Case Against Octopus Farming." *Issues in Science and Technology* XXXV, no. 2 (2019): 37–44.

82 ***hatchlings adapt to crowded quarters, as long as individuals are all of similar size:***
There has been a little published on rearing captive octopuses at high densities:

Rosas, C., Mascaró, M., Mena, R., Caamal-Monsreal, C. and Domingues, P.
"Effects of Different Prey and Rearing Densities on Growth and Survival of *Octopus
maya* Hatchlings." *Fisheries and Aquaculture Journal* 5, no. 4 (2014): 1–7.

Roura, Á., Martínez-Pérez, M., and Catro-Bugallo, A. "*Octopus vulgaris* Life
Cycle: Growth, Reproduction and Senescence in Captivity." Abstracts, 2022 CIAC
Symposium, Sesimbra, Portugal, April 4–8, 2022.

83 ***overall higher value from their octopus harvests:*** Oliver, T. A., Oleson, K. L., Ratsim-
bazafy, H., Raberinary, D., Benbow, S., and Harris, A. "Positive Catch & Economic
Benefits of Periodic Octopus Fishery Closures: Do Effective, Narrowly Targeted
Actions 'Catalyze' Broader Management?" *PLOS One* 10, no. 6 (2015): e0129075.

## CHAPTER 8: OCTOPUS SCRAPS

90 ***Why was the world's largest octopus species so often choosing these small crabs?:*** The
discussion of octopus diet choices draws on papers published over the span of the
study, including:

Marsteller, C. "Prey Availability and Preference of Giant Pacific Octopus (*Enteroctopus dofleini*) in Prince William Sound, Alaska." Alaska Pacific University, Anchorage, AK. Available from ProQuest Dissertations & Theses Global (2018): 2039990215.

Scheel, D., and Anderson, R. C. "Variability in the Diet Specialization of *Enteroctopus dofleini* in the Eastern Pacific Examined from Midden Contents." *American Malacological Bulletin* 30, no. 2 (2012): 1–13.

Scheel, D., Lauster, A., and Vincent, T. L. S. "Habitat Ecology of *Enteroctopus dofleini* from Middens and Live Prey Surveys in Prince William Sound, AK. In *Cephalopods Present and Past: New Insights and Fresh Perspectives*. Eds. Landman, N. H., Davis, R. A., and Mapes, R. H. (Berlin: Springer, 2007): 434–58.

Scheel, D., Leite, T. S., Mather, D. L., and Langford, K. "Diversity in the Diet of the Predator *Octopus cyanea* in the Coral Reef System of Moorea, French Polynesia." *Journal of Natural History* 51, no. 43–44 (2017): 2615–33.

Vincent, T. L. S., Scheel, D., and Hough, K. R. "Some Aspects of Diet and Foraging Behavior of *Octopus dofleini* (Wülker, 1910) in its Northernmost Range." *Marine Ecology* 19, no. 1 (1998): 13–29.

90 ***hunter-gather humans in Alaska were not picky eaters:*** The phrase "omnivore's dilemma" was coined and applied to the difficulties in human food selection in:

Rozin, Paul. "The Selection of Foods by Rats, Humans, and Other Animals." *Advances in the Study of Behavior* 6 (1976): 21–76.

The diversity of diet among the Unangax̂ was reported in:

Travis, J. "No Omnivore's Dilemma for Alaskan Hunter-Gatherers." *Science* 335 (2012): 898.

Travis was a science journalist reporting the results of the study as presented at a symposium prior to peer-reviewed publication.

90 ***The Aleutian Islands are the traditional lands of the Unangax̂ people:*** The Unangax̂ people also refer to themselves as Aleut, a name applied to them after first contact with Russians. For basic information about Unangax̂ culture identity, see the websites of the Qawalangin Tribe of Unalaska at https://www.qawalangin.com/unangax; and The Aleutian Pribilof Islands Association at https://www.apiai.org/departments/cultural-heritage-department/culture-history/history/.

93 ***grows an internal rootlike system of ramlets (the interna):*** The Rhizocephalan parasite of the black-clawed crab (*Lophopanopeus bellus*) was identified (for North Pacific water by Hildbrand Boschma in 1953 in "The Rhizocephala of the Pacific") as *Loxothylacus panopaei*, a species of parasite invasive along the US Atlantic coast. However,

that now is considered incorrect, as the Pacific side species is not the same as found in Atlantic waters. According to the Smithsonian Nemesis database, *Loxothylacus pano-paei*, even as an invasive species, is restricted to Atlantic waters: https://invasions.si .edu/nemesis/species_summary/89752.

In my summary of Rhizocephalan barnacle parasites, I benefited from reading:

Walker, Graham. "Introduction to the Rhizocephala (Crustacea: Cirripedia)." *Journal of Morphology* 249, no.1 (2001): 1–8.

95 *a single meal or foraging bout, or at most a few:* We only recorded fresh remains, deposited within a few days of each survey. When the octopus disposes of claws and carapaces at the den, these discards settle on the gravel berm into a midden. There the remains age, or lighter items may be dispersed by waves or currents. The new remains attract a film of diatoms and algae over the span of a few days, and later encrusting barnacles, worms, and other marine life. In this way, we could record only the remains with clean inner surfaces—those on the midden for too brief a time for colonization by even the quickest visible growths of diatoms or algae.

96 *enveloped the bird:* The British Columbia octopus was almost certainly *Enterocto-pus dofleini*; the West Atlantic octopus was a new species, later described as *Octopus insularis*.

Nightingale, A. "Who's Up for Lunch? A Gull-Eating Octopus in Victoria, BC." In *BirdFellow*. Vol. 2012. April 27, 2012.

Sazima, I., and de Almeida, L. B. "The Bird Kraken: Octopus Preys on a Sea Bird at an Oceanic Island in the Tropical West Atlantic." *Marine Biodiversity Records* 1, no. e47 (2008): 1–3.

## CHAPTER 9: OCTOPUS TOOLS

100 *He picked bidarkis:* A bidarki is the black leather or katy chiton, an intertidal mollusk collected for food. A similar word, *baidarka*, refers to an Aleut-style kayak.

102 *Tooth wear, however, is a strong indicator of age in sea otters and likely limits their life spans:* Sea otters live to thirteen to seventeen years. Nicholson, Teri E., et al. "Robust Age Estimation of Southern Sea Otters from Multiple Morphometrics." *Ecology and Evolution* 10, no. 16 (2020): 8592–609.

103 *230 suckers per arm:* Brewer, R. S. "A Unique Case of Bilateral Hectocotylization in the North Pacific Giant Octopus (*Enteroctopus dofleini*)." *Malacologia* 56, no. 1, 2 (2013): 297–300.

103 *These complexities allow the muscular action of the suction cup to draw a strong*

***vacuum on the water inside the cup:*** Recent work continues to inform our understanding of how octopus suction cups work in the water, and enables the development of robotic actuators that can grab surfaces underwater (or in medical fluids) using octopus-inspired designs.

Kier, W. M., and Smith, A. M. "The Structure and Adhesive Mechanism of Octopus Suckers." *Integrated Comparative Biology* 42 (2002): 1146–53.

Tramacere, F., Appel, E., Mazzolai, B., and Gorb, S. N. "Hairy Suckers: The Surface Microstructure and Its Possible Functional Significance in the *Octopus vulgaris* Sucker." *Beilstein Journal of Nanotechnology* 5, no. 1 (2014): 561–65.

Tramacere, F., Pugno, N. M., Kuba, M. J., and Mazzolai, B. "Unveiling the Morphology of the Acetabulum in Octopus Suckers and Its Role in Attachment." *Interface Focus* 5, no. 1 (2015): 20140050.

Baik, S., Park, Y., Lee, T.-J., Bhang, S. H., and Pang, C. "A Wet-Tolerant Adhesive Patch Inspired by Protuberances in Suction Cups of Octopi." *Nature* 546, no. 7658 (2017): 396–400.

Greco, G., Bosia, F., Tramacere, F., Mazzolai, B., and Pugno, N. M. "The Role of Hairs in the Adhesion of Octopus Suckers: A Hierarchical Peeling Approach." *Bioinspiration & Biomimetics* 15, no. 3 (2020): 035006.

106   ***the bolsters can direct the pressure of a bend in the radula ribbon:*** Messenger, J., and Young, J. Z. "The Radular Apparatus of Cephalopods." *Philosophical Transactions of the Royal Society of London. Series B: Biological Sciences* 354, no. 1380 (1999): 161–82.

107   ***bite through a leg to open them, leaving a telltale mark:*** Our team described these bite marks on crab legs in a publication about the field signs octopuses leave. The paleontological community, to my surprise, often cites this paper to understand marks left on fossilized remains, a utility I had not considered when writing this article.

Dodge, R., and Scheel, D. "Remains of the Prey—Recognizing the Midden Piles of *Octopus dofleini* (Wülker)." *The Veliger* 42 no. 3 (1999): 260–66.

## CHAPTER 10: STORIED OCTOPUSES

112   ***"In the story," I said, drawing a deep breath, "there is a hunter":*** "Chapter 47: Kanayĝaax̂tux̂" (from Timofey Dorofeyev, Umnak, 1909). In Bergsland, K., and Dirks, M. L. (Eds.). *Unangam Ungiikangin kayux Tunusangin, Aleut Tales and Narratives.* Collected 1909–1910 by Waldemar Jochelson (Fairbanks: Alaska Native Language Center, University of Alaska Fairbanks, 1990), 346–49.

116 *Moles and water shrews use the same underwater sniffing method when foraging:* Marriott, S., Cowan, E., Cohen, J., and Hallock, R. M. "Somatosensation, Echolocation, and Underwater Sniffing: Adaptations Allow Mammals without Traditional Olfactory Capabilities to Forage for Food Underwater." *Zoological Science* 30, no. 2 (2013): 69–75.

117 *no longer develop olfactory parts of the nervous system:* Kishida, Takushi. "Olfaction of Aquatic Amniotes." *Cell and Tissue Research* (2021): 1–13.

118 *acute vision underwater:* Strobel, Sarah McKay, et al. "Adaptations for Amphibious Vision in Sea Otters (*Enhydra lutris*): Structural and Functional Observations." *Journal of Comparative Physiology A* 206, no. 5 (2020): 767–82.

118 *the octopuses' first line of defense against visual predation:* The discussion was influenced by Mather, J. A., and Kuba, M. J. "The Cephalopod Specialties: Complex Nervous System, Learning, and Cognition." *Canadian Journal of Zoology* 91, no. 6 (2013): 431–49.

119 *disguised by posture and shape as a conspicuous coral head or an algae stalk like those surrounding it:* Josef, N., Amodio, P., Fiorito, G., and Shashar, N. "Camouflaging in a Complex Environment—Octopuses Use Specific Features of Their Surroundings for Background Matching." *PLOS One* 7, no. 5 (2012): e37579.

Josef, N., and Shashar, N. "Camouflage in Benthic Cephalopods: What Does It Teach Us?" In *Cephalopod Cognition* (Eds. Darmaillacq, A-S, Dickel, L., and Mather, J.) (Cambridge: Cambridge University Press, 2014), 177–96.

119 *Visual animals track moving things more easily than still ones:* Sam Kean discusses how this works neurologically in *The Tale of the Dueling Neurosurgeons: The History of the Human Brain as Revealed by True Stories of Trauma, Madness, and Recovery* (Boston: Little, Brown and Company, 2014.) The book follows the history of neuroscience through the brain injuries that served as early foundations of the science.

121 *the day octopus changes on average three times per minute:* Statistics are from Hanlon et al. (1999). For further reading about dynamic mimicry, see:

Hanlon, R. "Cephalopod Dynamic Camouflage." *Current Biology* 17, no. 11 (2007): R400–4.

Hanlon, R. T., Conroy, L.-A., and Forsythe, J. W. "Mimicry and Foraging Behaviour of Two Tropical Sand-Flat Octopus Species off North Sulawesi, Indonesia." *Biological Journal of the Linnean Society* 93, no. 1 (2008): 23–38.

Hanlon, R. T., Forsythe, J. W., and Joneschild, D. E. "Crypsis, Conspicuousness, Mimicry and Polyphenism as Antipredator Defences of Foraging Octopuses on Indo-

Pacific Coral Reefs, with a Method of Quantifying Crypsis from Video Tapes." *Biological Journal of the Linnean Society* 66, no. 1 (1999): 1–22.

Huffard, C. L., Saarman, N., Hamilton, H., and Simison, W. B. "The Evolution of Conspicuous Facultative Mimicry in Octopuses: An Example of Secondary Adaptation?" *Biological Journal of the Linnean Society* 101, no. 1 (2010): 68–77.

Krajewski, J. P., Bonaldo, R. M., Sazima, C., and Sazima, I. "Octopus Mimicking Its Follower Reef Fish. *Journal of Natural History* 43, nos. 3–4 (2009): 185–90.

Norman, M. D., Finn, J., and Tregenze, T. "Dynamic Mimicry in an Indo-Malayan Octopus." *Proceedings Royal Society of London* B 268 (2001): 1755–58.

Norman, M. D., and Hochberg, F. G. "The 'Mimic Octopus' (*Thaumoctopus mimicus* n. gen. et sp.), a New Octopus from the Tropical Indo-West Pacific (Cephalopoda: Octopodidae)." *Molluscan Research* 25, no. 2 (2006): 57–70.

121 ***Even three choices can cut the number of detections in half:*** Karpestam, E., Merilaita, S., and Forsman, A. "Natural Levels of Colour Polymorphism Reduce Performance of Visual Predators Searching for Camouflaged Prey." *Biological Journal of the Linnean Society* 112, no. 3 (2014): 546–55.

CHAPTER 11: OCTOPUS ADEPT

124 ***Octopus ink is composed of melanin and mucus:*** See these articles.

Bush, Stephanie L., and Bruce H. Robison. "Ink Utilization by Mesopelagic Squid." *Marine Biology* 152, no. 3 (2007): 485–94.

Caldwell, Roy L. "An Observation of Inking Behavior Protecting Adult *Octopus bocki* from Predation by Green Turtle (*Chelonia mydas*) Hatchlings." *Pacific Science* 59, no. 1 (2005): 69–72.

Derby, Charles D. "Cephalopod Ink: Production, Chemistry, Functions and Applications." *Marine Drugs* 12, no. 5 (2014): 2700–30.

Kelley, Jennifer L., and Wayne I. L. Davies. "The Biological Mechanisms and Behavioral Functions of Opsin-Based Light Detection by the Skin." *Frontiers in Ecology and Evolution* 4 (2016): 106.

Strugnell, Jan M., et al. "The Ink Sac Clouds Octopod Evolutionary History." *Hydrobiologia* 725, no. 1 (2014): 215–35.

126 ***recorded by snorkelers in Hanauma Bay:*** I based my account of this moray attack in the imagined life of the day octopus in Hawaii on a video of just such an encounter, filmed in Hanauma Bay, Oahu. The moray in this video appears to be struggling

with the octopus as much as the octopus is with the eel, as I describe in the text. The video may be found online at the National Geographic website: https://www .nationalgeographic.com/animals/article/eel-vs-octopus-video-hanauma-bay-hawaii.

127  *two types previously not found in the most comprehensive knot-tying manuals:* Previously undescribed knots tied by the Green Moray (*Gymnothorax prasinus*) in Australian waters included the Moray knot and the Moray banana knot. Both have a number of body crossings, making them bulkier with more potential for the moray to apply tearing leverage.

    Malcolm, Hamish A. "A Moray's Many Knots: Knot Tying Behaviour around Bait in Two species of Gymnothorax Moray Eel." *Environmental Biology of Fishes* 99, no. 12(2016): 939–47.

128  *reduce the size of the wound during the first few hours post-injury:* Wounds may be 80 percent covered within six hours of injury among fast-healing octopuses, although coverage may be only 50 percent or so among slow-healing octopuses. Much of my discussion of octopus wounds relied on:

    Imperadore, P., and Fiorito, G. "Cephalopod Tissue Regeneration: Consolidating over a Century of Knowledge." *Frontiers in Physiology* 9 (2018): 593.

    Shaw, T. J., Osborne, M., Ponte, G., Fiorito, G., and Andrews, P. L. "Mechanisms of Wound Closure Following Acute Arm Injury in *Octopus vulgaris.*" *Zoological Letters* 2, no. 8 (2016).

128  *the arms are sensitive to light:* Katz, Itamar, Shomrat, Tal, and Nesher, Nir. "Feel the Light: Sight-Independent Negative Phototactic Response in Octopus Arms." *Journal of Experimental Biology* 224, no. 5 (2021).

## CHAPTER 12: SEEING OCTOPUSES

137  *holds in view the direction of movement in the center of the fovea of the leading eye:* Levy, G., and Hochner, B. 2017. "Embodied Organization of *Octopus vulgaris* Morphology, Vision, and Locomotion." *Frontiers in Physiology* 8, 164.

138  *using either filters to convey color information . . . or . . . chromatic aberration:* Mäthger et al. (2010) offered a hypothesis that light receptive molecules might be closely associated with chromatophores in the skin. The chromatophores, acting as spectral filters, might make available color information to the animal. Stubbs and Stubbs (2016) suggest the chromatic aberration hypothesis.

    There are a number of studies confirming that the skin of cephalopods is light sensitive.

Buresch, K. C., Ulmer, K. M., Akkaynak, D., Allen, J. J., Mäthger, L. M., Nakamura, M., and Hanlon, R. T. "Cuttlefish Adjust Body Pattern Intensity with Respect to Substrate Intensity to Aid Camouflage, but Do Not Camouflage in Extremely Low Light." *Journal of Experimental Marine Biology and Ecology* 462 (2015): 121–26.

Katz, I., Shomrat, T., and Nesher, N. "Feel the Light: Sight-Independent Negative Phototactic Response in Octopus Arms." *Journal of Experimental Biology* 224, no. 5 (2021).

Mäthger, L. M., Roberts, S. B., and Hanlon R. T. "Evidence for Distributed Light Sensing in the Skin of Cuttlefish, *Sepia officinalis*." *Biology Letters* 6, no. 5 (2010): 600–3.

Ramirez, M. D., and Oakley, T. H. "Eye-Independent, Light-Activated Chromato-phore Expansion (LACE) and Expression of Phototransduction Genes in the Skin of *Octopus bimaculoides*." *Journal of Experimental Biology* 218, no. 10 (2015): 1513–20.

Stubbs, A. L., and Stubbs, C. W. "Spectral Discrimination in Color Blind Animals via Chromatic Aberration and Pupil Shape." *Proceedings of the National Academy of Sciences* 113, no. 29 (2016): 8206–11.

Al-Soudy, A.-S., Maselli, V., Galdiero, S., Kuba, M. J., Polese, G., and Di Cosmo, A. "Identification and Characterization of a Rhodopsin Kinase Gene in the Suckers of *Octopus vulgaris*: Looking around Using Arms?" *Biology* 10, no. 9 (2021): 936.

139 **people have tried to find such behavior:**

Messenger, J. B. "Evidence That Octopus Is Colour Blind." *Journal of Experimental Biology* 70 (1977): 49–55.

Mäthger, L. M., Barbosa, A., Miner, S., and Hanlon, R. T. "Color Blindness and Contrast Perception in Cuttlefish (*Sepia officinalis*) Determined by a Visual Senso-rimotor Assay." *Vision Research* 46, no. 11 (2006): 1746–53.

139 **most cephalopods lack:** The firefly squid (*Watasenia scintillans*) has three visual pigments, and probably sees color. This is a bioluminescent open-ocean squid that normally lives below two hundred meters depth, except during the spawning season when it is found in surface waters. Perhaps its color vision enables it to distinguish between natural light and its bioluminescent conspecifics. Three other mesopelagic nocturnal squids and the midwater gelatinous octopods *Japetella sp.* also have multiple visual pigments, again suggesting that color vision, where it occurs among cephalopods, may serve to detect bioluminescence.

Hanlon, R. T., and Messenger, J. B. *Cephalopod Behaviour*. 2nd edition. (Cambridge: Cambridge University Press, 2008).

141 ***the only animals known to use a concave mirror:*** Speiser, D. I., Loew, E. R., and Johnsen, S. "Spectral Sensitivity of the Concave Mirror Eyes of Scallops: Potential Influences of Habitat, Self-Screening and Longitudinal Chromatic Aberration." *Journal of Experimental Biology* 214, no. 3 (2011): 422–31.

141 ***Other aspects of scallop eyes:*** I referenced the following citations for this account:

Hamilton, P. V., and Koch, K. M. "Orientation toward Natural and Artificial Grassbeds by Swimming Bay Scallops, *Argopecten irradians* (Lamarck, 1819)." *Journal of Experimental Marine Biology and Ecology* 199, no. 1 (1996): 79–88.

Land, M. F. "Eyes with Mirror Optics." *Journal of Optics A: Pure and Applied Optics* 2, no. 6 (2000): R44.

Wilkens, L. A. "Chapter 5: Neurobiology and Behaviour of the Scallop." In *Developments in Aquaculture and Fisheries Science.* Vol. 35. Eds. S. E. Shumway and G. J. Parsons (Amsterdam: Elsevier, 2006), 317–56.

Palmer, B. A., Taylor, G. J., Brumfeld, V., Gur, D., Shemesh, M., Elad, N., Osherov, A., Oron, D., Weiner, S., and Addadi, L. "The Image-Forming Mirror in the Eye of the Scallop." *Science,* 358, no. 6367 (2017): 1172–75.

Miller, H. V., Kingston, A. C., Gagnon, Y. L., and Speiser, D. I. "The Mirror-Based Eyes of Scallops Demonstrate a Light-Evoked Pupillary Response." *Current Biology,* 29, no. 9 (2019): R313–14.

142 ***little to no binocular overlap of the visual range of the two eyes:*** Hanke, F. D., and Kelber, A. "The Eye of the Common Octopus (*Octopus vulgaris*)." *Frontiers in Physiology* 10 (2020): 1637.

143 ***the advantages of planning based on vision are at their peak:*** Mugan, U., and MacIver, M. A. "Spatial Planning with Long Visual Range Benefits Escape from Visual Predators in Complex Naturalistic Environments." *Nature Communications* 11, no. 1 (2020): 1–14.

Fitzgibbon, C. D. "A Cost to Individuals with Reduced Vigilance in Groups of Thomson's Gazelles Hunted by Cheetahs." *Animal Behaviour* 37, no. 3 (1989): 508–10.

Scheel, D. "Watching for Lions in the Grass: The Usefulness of Scanning and Its Effects during Hunts." *Animal Behaviour* 46, no. 4 (1993): 695–704.

## CHAPTER 13: REACHING OCTOPUSES

146 ***the words of Victor Hugo:*** Hugo, Victor. *Toilers of the Sea.* Ed. Ernest Rhys, Trans. W. Moy Thomas. The Project Gutenberg Ebook, 2010, retrieved April 22, 2022: https://www.gutenberg.org/files/32338/32338-h/32338-h.htm.

146 ***August 1903, off Victoria, BC:*** Details of this story appeared in *The Kendrick* gazette (Kendrick, Idaho), March 18, 1904. Chronicling America: Historic American Newspapers. Library of Congress: https://chroniclingamerica.loc.gov/lccn/sn86091096/1904-03-18/ed-1/seq-3.

147 ***I heard stories:*** A version of the blueberry story appears as "Woman and Octopus." In Krauss, M .E. (ed.). *In Honor of Eyak: The Art of Anna Nelson Harry* (Fairbanks: Alaska Native Language Center, University of Alaska Fairbanks, 1982).

> In this story, an Eyak woman with her child were picking blueberries, when she stumbled over an octopus in the bushes. "Long-fingers," she asked the octopus, "what are you doing here?" The octopus wrapped itself around her and abducted her to its undersea home. There the octopus married her. Her octopus husband provided well for her. He gathered many foods, lying over them to cook them, including seals, cockles and fish. In time, as the octopus-wife climbed up out of the water to rest on a skerry one day, the child's uncles happen by and entreated her to leave her octopus-home and return with them to land. She said "Let me stay here a little while yet; your octopus-brother-in-law will provide for you too," and her brothers relented.
>
> The Eyak woman lived among the octopuses, raising an octopus family, until in time she and her octopus family went ashore to live as humans. Out to sea one day fishing, her husband-octopus made a mistake and was killed by a whale. The woman lived with her sisters for a while before dying; her children returned to the sea to avenge their father and human family, but after killing the whale, they stayed in the sea and nothing more is known of them.

Octopuses coming ashore with agile suckers has long been part of human stories. In a five-book didactic epic poem on fisheries, Oppianus of Corycus (176–180 AD), a poet in ancient Rome, relates the love of the octopus for trees that flourished with splendid olives on the slope above the shore, to which octopuses are drawn from the water in close embrace before returning to the sea. Greek fishermen take advantage of this love, dragging olive branches through the water to which besotted octopuses are bound by desire and are thus caught.

*Oppian, Colluthus, and Tryphiodorus.* Halieutica (London: W. Heinemann, 1928), and (New York: G. P. Putnam's Sons). As described in Sauer et al. "World Octopus Fisheries." *Reviews in Fisheries Science & Aquaculture* (2019): 1–151.

I was told the gem about the knife attack by fisherman and raconteur Vern Robbins many years after he survived his stab wound. Walter McGregor of the Sealiontown people told a story of a man being dragged below the water to live with the octopuses, related in:

Swanton, J. R. *Haida Texts and Myths, Skidegate Dialect Recorded by John R. Swanton*. Smithsonian Institution Bureau of American Ethnology Bulletin 29 (Washington, DC: US Government Printing Office, 1905.) Reprinted (Brighton, MI: Native American Book Publishers, 1991).

148   *so many neurons outside the brain?:* My discussion of numbers and distributions of neurons in vertebrates, humans, and octopuses were drawn from the following:

Budelmann, B. U. "The Cephalopod Nervous System: What Evolution Has Made of the Molluscan Design." In *The Nervous Systems of Invertebrates: An Evolutionary and Comparative Approach*. Eds. Breidbach, O., and Kutsch, W. (Basel, Switzerland: Birkhauser Verlag, 1995), 115–38.

Herculano-Houzel, S. "Remarkable, But Not Special: What Human Brains Are Made of." *Evolutionary Neuroscience* (2020): 803–13.

Herculano-Houzel, Suzana. "The Human Brain in Numbers: A Linearly Scaled-Up Primate Brain." *Frontiers in Human Neuroscience* 3 (2009): 31.

News coverage of Herculano-Houzel's work: https://www.verywellmind.com/how-many-neurons-are-in-the-brain-2794889, as well as correspondence with Eric Chudler, PhD, University of Washington.

Furness, J. B., Callaghan, B. P., Rivera, L. R., Cho, H. J. "The Enteric Nervous System and Gastrointestinal Innervation: Integrated Local and Central Control." *Advances in Experimental Medicine and Biology* 817 (2014): 39–71. doi: 10.1007/978-1-4939-0897-4_3. PMID: 24997029.

148   *the octopus's neurological challenge arising from the lack of a hard skeleton:* This discussion of muscular hydrostats, octopus arm positioning, and propagation of waves down the arm is drawn from the following:

Gutfreund, Y., Flash, T., Yarom, Y., Fiorito, G., Segev, I., and Hochner B. "Organization of Octopus Arm Movements: A Model System for Studying the Control of Flexible Arms." *Journal of Neuroscience* 16, no. 22 (November 15, 1996): 7297–307.

Sumbre, G., Gutfreund, Y., Fiorito, G., Flash, T., and Hochner, B. "Control of Octopus Arm Extension by a Peripheral Motor Program." *Science* 293 (2001): 1845–48.

Sumbre, G., Fiorito, G., Flash, T., and Hochner, B. "Neurobiology: Motor Control of Flexible Octopus Arms." *Nature* 433, no. 7026 (2005): 595.

———. "Octopuses Use a Human-Like Strategy to Control Precise Point-to-Point Arm Movements." *Current Biology* 16, (2006): 767–72.

151 **Apollo Theater dancer Bill Bailey's "moonwalk":** The moonwalk dance move was already a fan favorite at the Apollo Theater, as performed in 1955 by tap dancer Bill Bailey. See it on YouTube at: https://www.youtube.com/watch?v=y71njpDH3co.

151 **octopuses crawl in any direction:** Levy, G., Flash, T., and Hochner, B. "Arm Coordination in Octopus Crawling Involves Unique Motor Control Strategies." *Current Biology* 25, no. 9 (2015): 1195–1200.

Kennedy, E. L., Buresch, K. C., Boinapally, P., and Hanlon, R. T. "Octopus Arms Exhibit Exceptional Flexibility." *Scientific Reports* 10, no. 1 (2020) 20872.

151 **Each arm tip stays within its own octant:** *Octant*, the eighth part of a circle, seems a perfect word for this application. The extent to which arm tips stay within their octants was shown during an investigation into octopus plume tracking that applies primarily to crawling octopuses. Likely this does not apply to walking octopuses that cross their third and fourth ipsilateral arms in quadrupedal gaits. The same study also provided arm positional data revealing the careful protection of the third right armtip by male octopuses.

Weertman, W., Scheel, D., Venkatesh, G., Gire, D. 2021. "An Investigation into Plume-guided Odor Search by Octopuses." Association for Chemoreception Sciences (AChemS) Virtual Meeting, April 19–23 2021.

152 **an octopus will try to envelop it in the web and arms:** An exception to this is the constriction hold, which relies on one arm. This will be described later (Chapter 18).

## CHAPTER 14: SENSATIONAL OCTOPUSES

156 **This is the story of how Raven filled the sea with life:** A story told by Dekinā'k!, of the Box-house people. In Swanton, J. R. *Tlingit Myths and Texts Recorded by John R. Swanton*. Smithsonian Institution, Bureau of American Ethnology Bulletin 39 (Washington, DC: US Government Printing Office, 1909.) Reprinted (Brighton, MI: Native American Book Publishers, 1990), 11.

157 **rubbing octopuses on the head or mantle with a lab-grade Q-tip:** An insight suggested by Sandy Talbot with the USGS genetics lab.

Hollenbeck, N., Scheel, D., Fowler, M., Sage, G. K., Toussaint, R. K., and Talbot, C. "Use of Swabs for Sampling Epithelial Cells for Molecular Genetics Analyses in *Enteroctopus*." *American Malacological Bulletin* 35, no. 2 (2017) 145–57.

Chancellor, S., Abbo, L., Grasse, B., Sakmar, T., Brown, J. S., Scheel, D., and Santymire, R. M. "Determining the Effectiveness of Using Dermal Swabs to Evaluate the Stress Physiology of Laboratory Cephalopods: A Preliminary Investigation." *General and Comparative Endocrinology* 314 (2021): 113903.

157 ***without any reference to the brain:*** The following sources were useful in writing this discussion of brain and sucker anatomy and function:

Budelmann, B. U. "The Cephalopod Nervous System: What Evolution Has Made of the Molluscan Design." In O. Breidbach and W. Kutsch, eds. *The Nervous Systems of Invertebrates: An Evolutionary and Comparative Approach* (Basel, Switzerland: Birkhauser Verlag, 1995), 115–38.

Fouke, K. E., and Rhodes, H. J. "Electrophysiological and Motor Responses to Chemosensory Stimuli in Isolated Cephalopod Arms." *The Biological Bulletin* 238, no. 1 (2020): 1–11.

Hague, T., Florini, M., and Andrews, P. L. "Preliminary In vitro Functional Evidence for Reflex Responses to Noxious Stimuli in the Arms of *Octopus vulgaris.*" *Journal of Experimental Marine Biology and Ecology* 447 (2013): 100–5.

Kier, W. M., and Smith, A. M. "The Morphology and Mechanics of Octopus Suckers." *The Biological Bulletin* 178 (1990): 126–36.

van Giesen, L., Kilian, P. B., Allard, C. A. H., and Bellano, N. W. "Molecular Basis of Chemotactile Sensation in Octopus." *Cell* 183 (2020): 594–604.

Young, J. Z. *A Model of the Brain.* (Oxford: Clarendon Press, 1964.)

157 ***They redirect their path:*** See Chapter 13, endnote 8 regarding Willem Weertman's plume-tracking study. Note that at the time of writing analyses are still underway and the interpretation preliminary.

159 ***one more tactic the scallop has to fool an octopus:*** The possibility that sponges provide tactile camouflage against octopus predation was informed by the following reports:

Bloom, S. A. "The Motile Escape Response of a Sessile Prey: A Sponge-Scallop Mutualism." *Journal of Experimental Marine Biology and Ecology* 17, 3 (1975): 311–21.

Forester, A. J. "The Association between the Sponge *Halichondria panicea* (Pallas) and Scallop *Chlamys varia* (L.): A Commensal-Protective Mutualism." *Journal of Experimental Marine Biology and Ecology* 36, no. 1 (1979): 1–10.

Pitcher, C., and Butler, A. "Predation by Asteroids, Escape Response, and Morphometrics of Scallops with Epizoic Sponges." *Journal of Experimental Marine Biology and Ecology* 112, no. 3 (1987): 233–49.

Harper, E. M. "Plio-Pleistocene Octopod Drilling Behavior in Scallops from Florida." *Palaios* 17, no. 3 (2002): 292–96.

Farren, H. M., and Donovan, D. A. "Effects of Sponge and Barnacle Encrustation on Survival of the Scallop *Chlamys hastata*." *Hydrobiologia* 592, no. 1 (2007): 225–34.

Onthank, K. L., Payne, B. G., Groeneweg, A. R., Ewing, T. J., Marsh, N., Cowles, D. L., and Lee, R. W. "A Sponge-Scallop Mutualism May Be Maintained by Octopus Predation." Chapter 3 in "Exploring the Life Histories of Cephalopods Using Stable Isotope Analysis of an Archival Tissue." PhD thesis by K. L. Onthank. Pullman, WA, Washington State University, 2013: 52–84.

160 **they prioritize chemical cues over visual in choosing food:** Maselli, V., Al-Soudy, A.-S., Buglione, M., Aria, M., Polese, G., and Di Cosmo, A. "Sensorial Hierarchy in *Octopus vulgaris*'s Food Choice: Chemical vs. Visual." *Animals* 10, no. 3 (2020): 457.

## CHAPTER 15: CONSTANT OCTOPUSES

167 **muscles coordinated by nervous systems:** The animal group Porifera, the sponges, has neither nerves nor muscles. The tiny animal *Trichoplax* is mobile, but lacking muscles it moves by ciliary action, and contains newly recognized gland cells that are likely neurosecretory and possibly minimal components constituting one necessary precursor to a nervous system. It is yet unknown whether *Trichoplax* fiber cells, which have similar characteristics to cells found in glass sponges and extend throughout the organism, have a neuron-like role in coordinating behavior.

Smith, Carolyn L., et al. "Novel Cell Types, Neurosecretory Cells, and Body Plan of the Early-Diverging Metazoan *Trichoplax adhaerens*." *Current Biology* 24, no. 14 (2014): 1565–72.

168 **distinguished the sizes of target squares despite varying distances from animal to target:** This study was:

Boycott, B., and Young, J. "Reactions to Shape in *Octopus vulgaris* Lamarck." Proceedings of the Zoological Society of London, 1956: 491–547.

This intriguing finding is barely mentioned in the paper. Its significance revealing perceptual constancy is discussed in Godfrey-Smith, P. *Other Minds: The Octopus, the Sea, and The Deep Origins of Consciousness* (New York: Farrar, Straus and Giroux, 2016), 84–100.

169 **a belt-and-compass device:** Kaspar, Kai, et al. "The Experience of New Sensorimotor Contingencies by Sensory Augmentation." *Consciousness and Cognition* 28 (2014): 47–63.

169 *from both the behaviors of the organism* and *from the environment:* In this section I was informed by Section 5.3 in van Woerkum, B. "Distributed Nervous System, Disunified Consciousness?: A Sensorimotor Integrationist Account of Octopus Consciousness." *Journal of Consciousness Studies* 27, nos. 1–2 (2020): 149–72.

170 *guided by the explicit formulation of a goal:* This section draws on a paper in which its authors discuss rigid behavioral programs as distinct from "intentions," that is, an outcome that the organism held in mind.

Scheel, D., Godfrey-Smith, P., Linquist, S., Chancellor, S., Hing, M., and Lawrence, M. "Octopus Engineering, Intentional and Inadvertent." *Communicative & Integrative Biology*, 11, no. 1 (2017). DOI: 10.1080/19420889.2017.1395994.

The gradations of complexity from inflexible responses to goal-formations were discussed by Sterelny, K. "Situated Agency and the Descent of Desire." In Hardcastle, V. G., ed. *Where Biology Meets Psychology: Philosophical Essays* (Cambridge, MA: MIT Press, 1999), 203–19.

170 *egg-rolling behavior of the nesting Greylag Geese:* Loren, K., and Tinbergen, N. "Taxis und Instinkthandlung in der Eirollbewegung der Graugans. I1." *Ethology* 2 (1939): 1–29.

171 *before a different behavior emerges:* For a fascinating discussion of both the variation among *Sphex* in this behavior, and the simplification of this example in philosophical literature, see:

Keijzer, Fred. "The Sphex Story: How the Cognitive Sciences Kept Repeating an Old and Questionable Anecdote." *Philosophical Psychology* 26, no. 4 (2013): 502–19.

However, diligent researchers have found further sphexishness in the behaviors of digger wasps. Females will provision burrows during the day strictly according to the results of their morning inspection, even if burrow contents have subsequently been switched by researchers, such that a still-attentive mother wasp might provision differently. But having completed inspection before the switch, the digger wasps do not attend to the change nor change their provisioning. See discussion, pp. 58–59, in Denton, D. *The Primordial Emotions: The Dawning of Consciousness* (Oxford: Oxford University Press, 2005).

The coinage of the term sphexishness appears in:

Hofstadter, D. R. "Metamagical Themas: Can Inspiration Be Mechanized?" *Scientific American* 247, no. 3 (1982): 18–34.

171 *According to one modeling study:* Mugan, U., and MacIver, M. A. "Spatial Planning

with Long Visual Range Benefits Escape from Visual Predators in Complex Natural-istic Environments." *Nature Communications* 11, 1 (2020): 1–14.

171 *cuttlefish raised in overly simplified environments:* Yasumuro, Haruhiko, and Yuzuru Ikeda. "Environmental Enrichment Affects the Ontogeny of Learning, Memory, and Depth Perception of the Pharaoh Cuttlefish *Sepia pharaonis.*" *Zoology* 128 (2018): 27–37.

173 *octopuses seem more unified than this:* A few studies support the generalization of both visual and tactile one-sided learning in octopuses.

Muntz, W. R. A. "Interocular Transfer in Octopus: Bilaterality of the Engram." *Journal of Comparative and Physiological Psychology* 54, no. 2 (1961): 192.

Wells, M. J., and Young, J. Z. "Lateral Interaction and Transfer in the Tactile Memory of the Octopus." *Journal of Experimental Biology* 45, no. 3 (1966): 383–400.

Wells, M. J. "Short-Term Learning and Interocular Transfer in Detour Experiments with Octopuses." *Journal of Experimental Biology* 47, no. 3 (1967): 393–408.

173 *octopuses use only limited proprioception information about the positions of their suckers:* Wells, M. "Tactile Discrimination of Surface Curvature and Shape by the Octopus." *Journal of Experimental Biology* 41 (1964): 433–45.

Grasso, F., and Wells, M. "Tactile Sensing in the Octopus." *Scholarpedia* 8, no. 6 (2013): 7165. http://www.scholarpedia.org/article/Tactile_sensing_in_the_octopus.

For further discussion of arm autonomy and octopus neurological organization, see:

Wells, M. J. *Octopus.* (London: Chapman and Hall, 1978.)

Young, J. Z. *The Anatomy of the Nervous System of* Octopus vulgaris (Oxford: Clarendon Press, 1971).

Nixon, M., and Young, J. Z. *The Brains and Lives of Cephalopods* (Oxford: Oxford University Press, 2003).

Gutnick, T., Zullo, L., Hochner, B., and Kuba, M. J. "Use of Peripheral Sensory Information for Central Nervous Control of Arm Movement by *Octopus vulgaris.*" *Current Biology* 30, no. 21 (2020): 4322–27.

173 *severed arms, when electrically stimulated, could perform whole-arm actions:* Sumbre, G., Gutfreund, Y., Fiorito, G., Flash, T., and Hochner, B. "Control of Octopus Arm Extension by a Peripheral Motor Program." *Science* 293 (2001): 1845–48.

173 *Perhaps the arms could learn without the brain:* Two considered works on the strange idea that octopus arms might be their own centers of learning or cognition are:

Grasso, F. W. "The Octopus with Two Brains: How Are Distributed and Central Representations Integrated in the Octopus Central Nervous System." In *Cephalopod Cognition*. Eds. Darmaillacq, A-S., Dickel, L., and Mather, J. (Cambridge: Cambridge University Press, 2014), 94–122.

Carls-Diamante, S. "Out on a Limb? On Multiple Cognitive Systems within the Octopus Nervous System." *Philosophical Psychology* (2019): 1–20.

174 ***octopuses can and do exert top-down control over their arms:*** See Gutnick et al. 2020, endnote 10, and:

Bagheri, H., Hu, A., Cummings, S., Roy, C., Casleton, R., Wan, A., Erjavic, N., Berman, S., Peet, M. M., and Aukes, D. M. "New Insights on the Control and Function of Octopus Suckers." *Advanced Intelligent Systems* (2020): 1900154.

174 ***cuttlefish can learn the same self-control that children demonstrate in a "marshmallow test":*** Outcomes for children in the marshmallow test are considered here:

Falk, Armin, Kosse, Fabian, and Pinger, Pia. "Re-Revisiting the Marshmallow Test: A Direct Comparison of Studies by Shoda, Mischel, and Peake (1990) and Watts, Duncan, and Quan" (2018). *Psychological Science* 31, no. 1 (2019): 100–4.

And for a cuttlefish equivalent delayed but preferred reward, here:

Schnell, A. K., Boeckle, M., Rivera, M., Clayton, N. S., and Hanlon, R. T. "Cuttlefish Exert Self-Control in a Delay of Gratification Task." *Proceedings of the Royal Society B* 288, no. 1946 (2021): 20203161.

## CHAPTER 16: DREAMING OCTOPUS

178 ***kept up all night by inquisitive scientists:*** Nath, R. D., Bedbrook, C. N., Abrams, M. J., Basinger, T., Bois, J. S., Prober, D. A., Sternberg, P. W., Gradinaru, V., and Goentoro, L. "The Jellyfish *Cassiopea* Exhibits a Sleep-Like State. *Current Biology* 27, no. 19 (2017): 2984–90. e3.

178 ***Octopuses got attention in the same timeframe:*** The cephalopod studies were:

Duntley, S. P., Uhles M., and Feren, S. "Sleep in the Cuttlefish *Sepia pharaonis*." *Sleep* 25 (2002): A159–A160.14.

Duntley, S. P., and Morrissey, M. J. "Sleep in the Cuttlefish *Sepia officinalis*." *Sleep* 26 (2003): A118.

———. "Sleep in the Cuttlefish." *Annals of Neurology* 56 (2004): S68.15.

Brown, E. R., Piscopo, S., De Stefano, R., and Giuditta, A. "Brain and Behavioural Evidence for Rest-Activity Cycles in *Octopus vulgaris*." *Behavioural Brain Research* 172, no. 2 (2006): 355–59.

Meisel, D. V., Byrne, R. A., Mather, J. A., and Kuba, M. "Behavioral Sleep in *Octopus vulgaris.*" *Vie et milieu / Life and Environment* 61, no. 4 (2011): 185–90.

Frank, M. G., Waldrop, R. H., Dumoulin, M., Aton, S., and Boal, J. G. "A Preliminary Analysis of Sleep-Like States in the Cuttlefish *Sepia officinalis.*" *PLOS One* 7, no. 6 (2012): e38125.

179 **to learn more about octopus sleep:** Medeiros, S. L. d. S., de Paiva, M. M. M., Lopes, P. H., Blanco, W., de Lima, F. D., de Oliveira, J. B. C., Medeiros, I. G., Sequerra, E. B., de Souza, S., Leite, T. S., and Ribeiro, S. "Cyclic Alternation of Quiet and Active Sleep States in the Octopus." *iScience* 24, no. 4 (2021): 102223.

179 **Do animals dream? And if so, can we know anything about it?:** This discussion arose during the making of BBC Natural World *The Octopus in My House* (2019–2020, also released in 2019 as PBS Nature: *Octopus Making Contact*) due to the work of Director of Photography Ernie Kovacs and Director Anna Fitch. You can view the video of Heidi dreaming online at https://www.pbs.org/wnet/nature/octopus-dreaming -trept6/19376/.

The material in this chapter is drawn from collaboration with Josie Malinowski and Mitchel McCloskey, and our paper "Do Animals Dream?" in *Consciousness & Cognition* 95 (2021): 103214.

My approach to the importance of animal dreaming was informed by a May 2022 book on the subject, *When Animals Dream: The Hidden World of Animal Consciousness* by David M. Peña-Guzmán (Princeton University Press). The book reviews the scientific evidence and philosophical perspective on the subject and concludes, I think justifiably, that, yes, many animals dream.

182 **replaying and also "hearing" the same patterns that produce song when awake:** Dave, Amish S., and Margoliash, Daniel. "Song Replay During Sleep and Computational Rules for Sensorimotor Vocal Learning." *Science* 290, no. 5492 (2000): 812–16.

182 **Alex rehearsed his new conversational skills in monologues:** Pepperberg, Irene M., Brese, Katherine J., and Harris, Barbara J. "Solitary Sound Play during Acquisition of English Vocalizations by an African Grey Parrot (*Psittacus erithacus*): Possible Parallels with Children's Monologue Speech." *Applied Psycholinguistics* 12, no. 2 (1991): 151–78.

183 **The brain's reward system is the same system that drives dreaming:** As described in Solms, Mark. *The Hidden Spring: A Journey to the Source of Consciousness* (New York: W. W. Norton & Company, 2021), 27–28.

CHAPTER 17: OCTOPUS HUNGRY AND AFRAID

191 **Derek Denton named these ancient and demanding urges the primordial emotions:** Denton, D. *The Primordial Emotions: The Dawning of Consciousness* (Oxford: Oxford University Press, 2005).

192 **triggered by stretch receptors in the crop:** Wells, M., and Wells, J. "Fluid Flow into the Gut of Octopus." *Vie et milieu / Life and Environment* (1988): 221–26.

Wells, M. J., and Wells, J. "Fluid uptake and the maintenance of blood volume in Octopus." *Journal of Experimental Biology* 175, no. 1 (1993): 211–18.

Fernández-Gago, R., Heß, M., Gensler, H., and Rocha, F. "3D Reconstruction of the Digestive System in *Octopus vulgaris* Cuvier, 1797 Embryos and Paralarvae during the First Month of Life." *Frontiers in Physiology* 8 (2017): 462.

192 **satisfy the sensation or relieve the imperative:** These drives compose the first of the three parts of the sense of self: motivation, concrete knowledge such as autobiographical memories, and abstract knowledge of self, according to Keenan *The Face in the Mirror* (New York: HarperCollins Publishers, 2003), p. 98, discussing Miller, B. L., Seeley, W. W., Mychack, P., Rosen, H. J., Mena, I., and Boone, K. "Neuroanatomy of the Self: Evidence from Patients with Frontotemporal Dementia." *Neurology* 57, no. 5 (2001): 817–21.

193 **expanding its mantle to as much as four times the volume of normal resting ventilations:** Wells, M. J., and Wells, J. "Ventilation Frequencies and Stroke Volumes in Acute Hypoxia in *Octopus.*" *Journal of Experimental Biology* 118 (1985): 445–48.

193 **an offered meal following thirty-six hours of fasting:** Short-term food deprivation makes octopuses more likely to participate in a researcher's experimental design, permitting some general observations about octopus motivation related to food. My discussion of hunger in this sections was based on reports of digestion and motivation from food deprivation:

Boucher-Rodoni, Renata, and Mangold, Katharina. "Experimental Study of Digestion in *Octopus vulgaris* (Cephalopoda: Octopoda)." *Journal of Zoology* 183, no. 4 (1977): 505–15.

Bastos, Penélope, et al. "Digestive Enzymes and Timing of Digestion in *Octopus vulgaris* Type II." *Aquaculture Reports* 16 (2020): 100262.

Linares, Marcela, et al. "Timing of Digestion, Absorption and Assimilation in Octopus Species from Tropical (*Octopus maya*) and Subtropical-Temperate (*O. mimus*) Ecosystems." *Aquatic Biology* 24, no. 2 (2015): 127–40.

Walker, J. J., Longo, N., and Bitterman, M. "The Octopus in the Laboratory. Handling, Maintenance, Training." *Behavior Research Methods & Instrumentation* 2, no. 1 (1970): 15–18.

Kuba, M. J., Byrne, R. A., Meisel, D. V., and Mather, J. A. "When Do Octopuses Play? Effects of Repeated Testing, Object Type, Age, and Food Deprivation on Object Play in *Octopus vulgaris*." *Journal of Comparative Psychology* 120, no. 3 (2006): 184.

———. "Exploration and Habituation in Intact Free Moving *Octopus vulgaris*." *International Journal of Comparative Psychology* 19 (2006): 426–38.

194 ***Octopuses settle down in a favorite spot:*** Meisel, D. V., Byrne, R. A., Mather, J. A., and Kuba, M. "Behavioral Sleep in *Octopus vulgaris*." *Vie et milieu / Life and Environment* 61, no. 4 (2011): 185–90.

Medeiros, S. L. d. S., de Paiva, M. M. M., Lopes, P. H., Blanco, W., de Lima, F. D., de Oliveira, J. B. C., Medeiros, I. G., Sequerra, E. B., de Souza, S., Leite, T. S., and Ribeiro, S. "Cyclic Alternation of Quiet and Active Sleep States in the Octopus." *iScience* (2021): https://doi.org/10.1016/j.isci.2021.102223.

195 ***do not adhere at all to octopus skin:*** Nesher, N., Levy, G., Grasso, F. W., and Hochner, B. "Self-Recognition Mechanism between Skin and Suckers Prevents Octopus Arms from Interfering with Each Other." *Current Biology* 24, no. 11 (2014): 1271–75.

## CHAPTER 18: OCTOPUS CANNIBAL

198 ***Octopuses eat one another:*** For a scientific account of the event opening this chapter and another similar one, see:

Hernández-Urcera, J., Garci, M. E., Roura, Á., González, Á. F., Cabanellas-Reboredo, M., Morales-Nin, B., and Guerra, Á. "Cannibalistic Behavior of Octopus (*Octopus vulgaris*) in the Wild." *Journal of Comparative Psychology* 128, no. 4 (2014): 427.

Hernández-Urcera, J., Cabanellas-Reboredo, M., Garci, M. E., Buchheim, J., Gross, S., Guerra, A., and Scheel, D. "Cannibalistic Attack by *Octopus vulgaris* in the Wild: Behaviour of Predator and Prey." *Journal of Molluscan Studies* 85, no. 3 (2019): 354–57.

199 ***cannibalism is common among the cephalopods, according to a 2010 publication:***
Ibáñez, C. M., and Keyl, F. "Cannibalism in Cephalopods." *Reviews in Fish Biology and Fisheries* 20, no. 1 (2010): 123–36.

199 ***no successful depredation of one octopus by another occurred in this study:*** The discussion of cannibalism in *Macroctopus maorum* drew from the following works. References to size differences appear in these articles and those cited in the previous two endnotes.

Anderson, T. J. "Morphology and Biology of *Octopus maorum* Hutton 1880 in Northern New Zealand." *Bulletin of Marine Science* 65, no. 3 (1999): 657–76.

Grubert, M. A., Wadley, V. A., and White, R. W. G. "Diet and Feeding Strategy of *Octopus maorum* in Southeast Tasmania." *Bulletin of Marine Science* 65, no. 2 (1999): 441–51.

200 **during a foraging study in Palau:** Hanlon, R. T., and Forsythe, J. W. "Sexual Cannibalism by *Octopus cyanea* on a Pacific Coral Reef." *Marine and Freshwater Behaviour and Physiology* 41, no. 1 (2008): 19–28.

201 **to restrain another octopus with a throttlehold:** Huffard, C. L., and Bartick, M. "Wild *Wunderpus photogenicus* and *Octopus cyanea* Employ Asphyxiating 'Constricting' in Interactions with Other Octopuses." *Molluscan Research* 35, no. 1 (2014): 12–16.

Scheel, D. "Octopuses in Wild and Domestic Relationships." *Social Science Information* 57, no. 3 (2018): 403–21.

## CHAPTER 19: OCTOPUSES IN WILD RELATIONSHIPS

203 **Southern Bastard Codlings:** Some of the ideas presented here were informed by discussions with Peter Godfrey-Smith and by the article "32. Housecats" on his blog *The Giant Cuttlefish*, http://giantcuttlefish.com/?p=3438.

204 **will have capacities for forming relationships between individuals:** Many of the ideas in this chapter were first developed in my 2018 paper and references therein. Scheel, D. "Octopuses in Wild and Domestic Relationships." *Social Science Information* 57, no. 3 (2018): 403–21.

205 **In a test with day octopuses:** Scheel, D., Leite, T. S., Mather, D. L., and Langford, K. "Diversity in the Diet of the Predator *Octopus cyanea* in the Coral Reef System of Moorea, French Polynesia." *Journal of Natural History* 51, nos. 43–44 (2017): 2615–33.

205 **specific areas of the brain are dedicated to the categorization of human faces:** Some bizarre effects are discussed by Oliver Sacks and more recently Sam Kean.

Sacks, O. *The Man Who Mistook His Wife for a Hat* (New York: Touchstone, 1998).

Kean, S. *The Tale of the Dueling Neurosurgeons: And Other True Stories of Trauma, Madness, Affliction, and Recovery That Reveal the Surprising History of the Human Brain* (Boston: Little, Brown and Company, 2014).

205 **traditional owners of these areas:** This information was drawn from several online sources:

Native Land Digital: https://native-land.ca/.

Map of Indigenous Australia: https://aiatsis.gov.au/explore/map-indigenous -australia.

Jervis Bay Wild: https://www.jervisbaywild.com.au/blog/brief-history-jervis-bay/.

Booderee National Park: https://parksaustralia.gov.au/booderee/.

207 ***a gloomy octopus cautiously crept into view inside its den:*** These observations were made from video captured August 6, 2013, from a south-placed camera following our third dive. Following this interaction, the light declined as the sun set.

209 ***attention from cleaning gobies:*** The accounts of octopuses with cleaning gobies were reported in:

Johnson, W. S., and Chase, V. C. "A Record of Cleaning Symbiosis Involving *Gobiosoma* sp. and a Large Caribbean Octopus." *Copeia* 1982, no. 3 (1982): 712–14.

209 ***Coral trout are drawn to foraging octopuses:*** Coral trout, groupers, goatfish, and other species may commonly follow foraging octopuses. I have photographed this behavior in Australia (coral trout) and Mo'orea (goatfish). The associations fascinate behavioral biologists, and a number of studies and anecdotal reports have been published describing them.

Bayley, D., and Rose, A. "Multi-Species Co-operative Hunting Behaviour in a Remote Indian Ocean Reef System." *Marine and Freshwater Behaviour and Physiology* 53, no. 1 (2020): 35–42.

Diamant, A., and Shpigel, M. "Interspecific Feeding Associations of Groupers (Teleostei: Serranidae) with Octopuses and Moray Eels in the Gulf of Eilat (Aqaba)." *Environmental Biology of Fishes* 13, no. 2 (1985): 153–59.

Krajewski, J. P., Bonaldo, R. M., Sazima, C., and Sazima, I. "Octopus Mimicking Its Follower Reef Fish." *Journal of Natural History* 43, nos. 3–4 (2009): 185–90.

Mather, J. A. "Interactions of Juvenile *Octopus vulgaris* with Scavenging and Territorial Fishes." *Marine Behaviour and Physiology* 19 (1992): 175–82.

Pereira, P. H. C., de Moraes, R. L. G., Feitosa, J. L. L., and Ferreira B. P. " 'Following the Leader': First Record of a Species from the Genus Lutjanus Acting as a Follower of an Octopus." *Marine Biodiversity Records* 4, no. 1 (2011).

Sazima, C., Krajewski, J. P., Bonaldo, R. M., and Sazima I. "Nuclear-Follower Foraging Associations of Reef Fishes and Other Animals at an Oceanic Archipelago." *Environmental Biology of Fishes* 80, no. 4 (2007): 351–61.

Strand, S. "Following Behavior: Interspecific Foraging Associations among Gulf of California Reef Fishes." *Copeia* (1988): 351–57.

Unsworth, R. K., and Cullen-Unsworth, L. C. "An Inter-specific Behavioural

Association between a Highfin Grouper (*Epinephelus maculatus*) and a Reef Octopus (*Octopus cyanea*)." *Marine Biodiversity Records* 5 (2012).

210 *it does occur in a few species:* Vail, A. L., Manica, A., and Bshary, R. "Referential Gestures in Fish Collaborative Hunting." *Nature Communications* 4 (2013): 1765.

Tauzin, Tibor, et al. "What or Where? The Meaning of Referential Human Pointing for Dogs (*Canis familiaris*)." *Journal of Comparative Psychology* 129, no. 4 (2015): 334.

Xitco, Mark J., Gory, John D., and Kuczaj, Stan A. "Spontaneous Pointing by Bottlenose Dolphins (*Tursiops truncatus*)." *Animal Cognition* 4, no. 2 (2001): 115–23.

## CHAPTER 20: GATHERING OCTOPUSES

220 **scientists published an account of their behavior:** Caldwell, R. L., Ross, R., Rodaniche, A., and Huffard, C. L. "Behavior and Body Patterns of the Larger Pacific Striped Octopus." *PLOS One* 10, no. 8 (2015): e0134152.

220 **lesser Pacific striped octopus (Octopus chierchiae):** Grearson, A. G., Dugan, A., Sakmar, T., Dölen, G., Gire, D. H., Sivitilli, D. M., Niell, C., Caldwell, R. L., Wang, Z. Y., and Grasse, B. "The Lesser Pacific Striped Octopus, *Octopus chierchiae*: An Emerging Laboratory Model for the Study of Octopuses." *bioRxiv* (2021).

Rodaniche, A. F. "Iteroparity in the Lesser Pacific Striped Octopus *chierchiae* (Jatta, 1889)." *Bulletin of Marine Science* 35, no. 1 (1984): 99–104.

221 **males and females in captivity will tolerate sharing:** Edsinger, E., Pnini, R., Ono, N., Yanagisawa, R., Dever, K., and Miller, J. "Social Tolerance in *Octopus laqueus*—A Maximum Entropy Model." *PLOS One* 15, no. 6 (2020): e0233834.

221 **involved in common encounters with each other:** Huffard, C. L. "Ethogram of *Abdopus aculeatus* (d'Orbigny, 1834) (Cephalopoda: Octopodidae): Can Behavioural Characters Inform Octopodid Taxomony and Systematics?" *Journal of Molluscan Studies* 73, no. 2 (2007): 185–93.

Huffard, C. L., Caldwell, R. L., and Boneka, F. "Male-Male and Male-Female Aggression May Influence Mating Associations in Wild Octopuses (*Abdopus aculeatus*)." *Journal of Comparative Psychology* 124, no. 1 (2010): 38–46.

221 **a female octopus tended her eggs for nearly four and a half years:** Robison, B., Seibel, B., and Drazen, J. "Deep-Sea Octopus (*Graneledone boreopacifica*) Conducts the Longest-Known Egg-Brooding Period of Any Animal." *PLOS One* 9, no. 7 (2014): e103437.

Drazen, J. C., Goffredi, S. K., Schlining, B., and Stakes, D. S. "Aggregations of Egg-Brooding Deep-Sea Fish and Cephalopods on the Gorda Escarpment: A Reproductive Hot Spot. *The Biological Bulletin* 205, no. 1 (2003): 1–7.

222 ***each egg-tending octopus within easy reach of neighbors:*** Raineault, N. A., Flanders, J., and Niiler, E., eds. "New Frontiers in Ocean Exploration: The E/V *Nautilus*, NOAA Ship *Okeanos Explorer*, and R/V *Falkor* 2020 Field Season." *Oceanography* 34, no. 1 (2021): supplement: https://doi.org/10.5670/oceanog.2021.supplement.01.

See also the award-winning short film *Discover Wonder: The Octopus Garden*. Sanctuary Integrated Monitoring Network: https://sanctuarysimon.org/2021/04/discover-wonder-the-octopus-garden-wins-best-short-film-at-the-international-ocean-film-festival/.

223 ***A gloomy octopus pushed at a shell:*** It is worth a note here about these vignettes that appear in each chapter from the lives of octopuses. All are based on actual moments I have observed or documented. The bouts of den cleaning, collecting the sponge, and interacting with another octopus are related from a single video sequence captured on January 16, 2016, from a north-placed camera following our second dive. In a few cases, I have taken minor liberties in reconstructing such moments: this chapter's opening vignette of the female foraging is assembled from several different observations and moments. Specific details, however, occurred as related.

224 ***the recreational drug ecstasy:*** Dölen, G., Darvishzadeh, A., Huang, K. W., and Malenka, R. C. "Social Reward Requires Coordinated Activity of Nucleus Accumbens Oxytocin and Serotonin." *Nature* 501, no. 7466 (2013): 179–84.

Edsinger, E., and Dölen, G. "A Conserved Role for Serotonergic Neurotransmission in Mediating Social Behavior in Octopus." *Current Biology* 28, no. 19 (2018): 3136–42. e4.

Dölen, G. "Mind Reading Emerged at Least Twice in the Course of Evolution. In *Think Tank: Forty Neuroscientists Explore the Biological Roots of Human Experience*. Ed. Linden, D. J. (New Haven, CT: Yale University Press, 2018), 194–200.

## CHAPTER 21: OCTOPUS QUIDNUNCS

225 ***another larger female rose up out of her den:*** This account is not from any single moment, but is composed of numerous instances across many video observations, as summarized, for example in scientific reports cited in Chapter 22, endnotes 4–6, *The Nosferatu is a signal*, page 295; *this low posture just described* , page 295; and *a female gathered up a bolus of silt and shells, and directed it through the water at a male*, page 296.

226 ***Males also are careful of their third right arm:*** Kennedy, E. L., Buresch, K. C., Boinapally, P., and Hanlon, R. T. "Octopus Arms Exhibit Exceptional Flexibility." *Scientific Reports* 10, no. 1 (2020): 20872.

227 *He is the octopus **Quidnunc:*** A Quidnunc is one who is curious to know everything that passes, and is continually asking "What now?" or "What news?" hence, one who knows or pretends to know all that is going on in politics, society, etc.; a newsmonger. From the Latin *quid nunc*, "what now." The Century Dictionary. http://www.global -language.com/CENTURY/.

228 *individually distinctive patterns of markings:* Huffard, C. L., Caldewell, R. L., DeLoach, N., Gentry, G. W., Humann, P., MacDonald, B., Moore, B., Ross, R., Uno, T., and Wong, S. "Individually Unique Body Patterns in Octopus (*Wunderpus photo-genicus*) Allow for Photoidentification." *PLOS One* 3, no. 11 (2008): e3732.

Caldwell, R. L., Ross, R., Rodaniche, A., and Huffard, C. L. "Behavior and Body Patterns of the Larger Pacific Striped Octopus." *PLOS One* 10, no. 8 (2015): e0134152.

228 *Damage or partial amputation of one or more of an octopus's arms are also common:* In one study of three different octopus species in California, more than half of the ani-mals had at least a portion of one or more arms missing. Some individuals had seven of eight arms injured. Octopuses generally are prone to damage to the front arm pairs.

Voss, K. M., and Mehta, R. S. "Asymmetry in the Frequency and Proportion of Arm Truncation in Three Sympatric California Octopus Species." *Zoology* (2021): 125940.

231 *the evidence for cephalopod recognition abilities is weak:* Boal, J. G. "Social Recogni-tion: A Top Down View of Cephalopod Behaviour." *Vie et milieu / Life and Environ-ment* 56, no. 2 (2006): 69–79.

232 *Paper wasps (**Polistes fuscatus***) recognize their nest-mates' faces:* These wasps have distinctive yellow markings on their black and brown faces. Familiar faces are greeted nonaggressively with body contact, but when facial patterns were manipulated, wasps lunged at the unfamiliar face with mandibles open, and were more likely to bite or mount the unfamiliar wasp.

Injaian, A., and Tibbetts, E. A. "Cognition across Castes: Individual Recognition in Worker *Polistes fuscatus* Wasps." *Animal Behaviour* 87 (2014): 91–96.

232 *Event memory entails a past moment of self:* Cognitive scientists also sometimes distinguish between *event* memory (a scene recalled as one occurrence) and *episodic* memory (an experientially relived scene about oneself recalled voluntarily from a sin-gle instance of memory formation). The stipulation of being "experientially relived," however, is challenging: absent a verbal report, there are no definitive behavioral markers of conscious experience. Animal behaviorists refer to *episodic-like* memory and, for many, this is the same as event memory. For our purposes, the key feature of

both is that remembering a particular scene (either as event or episode) entails a viewpoint of the self, located in space and time.

The use of behavioral evidence to establish that an animal recalls lived experiences is scientifically and philosophically controversial. Much of the controversy arises from the so-called zombie problem, or the "hard problem" of consciousness (e.g., Chalmers 1999, Nagel 1974). The zombie problem imagines that there could be a person, identical to me in all discernible ways and behaviors under all circumstances, who has no inner life, and feels nothing (though it reacts as if he does): an experiential zombie. The hard problem of consciousness is roughly the same idea in more general form—consciousness can be imagined as an unnecessary add-on to any complex responsive system, which could (we imagine) react the same way without awareness.

Such perspectives on consciousness have been thoroughly considered and rebutted by others, particularly Daniel Dennett (e.g., Dennett 1993, 2008). The views I follow here adopt an understanding of consciousness rooted in evolutionary theory, and accessible to science (e.g., Denton 2005, Godfrey-Smith 2016, 2020), albeit with challenges.

Clayton, Nicola S., et al. "Elements of Episodic-Like Memory in Animals." *Philosophical Transactions of the Royal Society of London. Series B: Biological Sciences* 356, no. 1413 (2001): 1483–91.

Dennett, Daniel. C. *Consciousness Explained* (London: Penguin UK, 1993).

Dennett, D. C. *Kinds of Minds: Toward an Understanding of Consciousness* (New York: Basic Books, 2008).

Denton, D. *The Primordial Emotions: The Dawning of Consciousness* (Oxford: Oxford University Press, 2005).

Godfrey-Smith, P. *Other Minds: The Octopus, the Sea, and the Deep Origins of Consciousness* (New York: Macmillan, 2016).

———. *Metazoa: Animal Life and the Birth of the Mind* (New York: Farrar, Straus and Giroux, 2020).

Nagel, Thomas. "What Is It Like to Be a Bat?" *Readings in Philosophy of Psychology* 1 (1974): 159–68.

Rubin, D. C., and Umanath S. "Event Memory: A Theory of Memory for Laboratory, Autobiographical, and Fictional Events." *Psychological Review* 122, no. 1 (2015): 1–23.

For criticism of this approach that behavior reveals event memory and complex cognition, see:

Malanowski, S. "Is Episodic Memory Uniquely Human? Evaluating the Episodic-Like Memory Research Program." *Synthese* 193, no. 5 (2016): 1433–55.

van Woerkum, Bas. "The Evolution of Episodic-Like Memory: The Importance of Biological and Ecological Constraints." *Biology & Philosophy* 36, no. 2 (2021): 1–18.

Schnell, A. K., et al. "How Intelligent Is a Cephalopod? Lessons from Comparative Cognition." *Biological Reviews* 96, no. 1 (2021): 162–78.

232 **Western scrub jays**: The proper accounting of the study subjects in this research requires knowing the changing taxonomy (as scientists have discovered that the various similar scrub jays are not all members of a single species), as well as the changing nomenclature (as old common names are retired and new ones added).

Two species were once lumped together under the common name Western scrub jay (which is no longer standard): the California Scrub Jay and Woodhouse's Scrub Jay. (Ornithologists capitalize standardized common bird names like proper nouns, whereas nonstandard names or those referring to groups of species are not capitalized.) The broader common term "scrub jay" encompassed Western scrub jays plus Island and Florida Scrub Jays. In the original report of episodic-like memory research (Clayton and Dickinson 1998), the study subjects are described as "scrub jays (*Aphelocoma coerulescens*)," the species name now belonging to the standardized common name Florida Scrub Jays. Neither author of the study listed a Florida affiliation. Salwiczek et al. (2010) summarize the same study as one of "memory in Western scrub-jays (*Aphelocoma californica*)"—note the common and species name differences. The common name of *A. californica* is now the California Scrub Jay.

Clayton, N. S., and Dickinson, A. "Episodic-Like Memory during Cache Recovery by Scrub Jays." *Nature* 395, no. 6699 (1998): 272–74.

Salwiczek, L. H., Watanabe, A., and Clayton, N. S. "Ten Years of Research into Avian Models of Episodic-Like Memory and Its Implications for Developmental and Comparative Cognition." *Behavioural Brain Research* 215, no. 2 (2010): 221–34.

233 **and a fish:** Hamilton, T. J., Myggland, A., Duperreault, E., May, Z., Gallup, J., Powell, R. A., Schalomon, M., and Digweed, S. M. "Episodic-Like Memory in Zebrafish." *Animal Cognition* 19, no. 6 (2016): 1071–79.

233 **Cuttlefish thus learned a what-where-when discrimination:** Jozet-Alves, C., Bertin, M., and Clayton N. S. "Evidence of Episodic-Like Memory in Cuttlefish." *Current Biology* 23, no. 23 (2013): R1033–35.

233 **decisions about how they would interact with us:** The "stay away" message of getting squirted by an octopus makes a good story. I have wondered, however, if octopuses have a different interest in squirting their keepers. Sometimes that first soaking from a large octopus is unexpected and gets a big response—a startled jump, shriek,

or laughter: something interesting happens. Are octopuses in captivity prone to squirt the most reactive people in the room, because it makes for interesting theater?

## CHAPTER 22: OCTOPUSES IN DOMESTIC RELATIONSHIPS

236 ***Octopuses are also notoriously solitary:*** In a simple laboratory test using the California two-spot octopus (*Octopus bimaculoides*), a species that Bateson had likely studied, adult female octopuses preferred to spend time near another female rather than to explore a novel object. Researchers sequencing octopus genomes found that the cephalopods have similar (but not identical) gene sequences to the human binding site for the psychoactive drug ecstasy (which overlaps the binding sites of the neurotransmitter serotonin). To test the function of these sequences, researchers gave ecstasy to octopuses. For humans, ecstasy stimulates neurotransmitters, in ways that increase self-confidence, sociability, and perceptions of shared empathy. Recreational users of ecstasy feel the urge to touch others. The researchers found that ecstasy increased the time the octopuses spent near to or touching another octopus, compared to time exploring a novel object. Attraction in animals has deep roots. The genetic and hormonal mechanisms for social attraction exist in octopuses. See Chapter 20, endnote 8 *the recreational drug ecstasy*, page 291.

237 ***Bateson then wrote a remarkable letter:*** My account of the letter and Bateson's interest in octopuses and the Cuban Missile crisis relied on Guddemi, P. *Gregory Bateson on Relational Communication: From Octopuses to Nations* (Berlin: Springer International Publishing, 2020).

238 ***few independent observations exist to expand on Bateson's account:*** Research on aquaculture of *Octopus maya* and *Octopus vulgaris* is beginning to put an end to this, as successful aquaculture will require understanding the extent to which octopuses can be housed together, as discussed briefly at the end of Chapter 7. Many challenges remain, however, even for octopuses found together in nature, as revealed by the difficulties of housing octopuses together as reported in Caldwell, R. L., Ross, R., Rodaniche, A., and Huffard, C. L. "Behavior and Body Patterns of the Larger Pacific Striped Octopus." *PLOS One* 10, no. 8 (2015): e0134152.

239 ***The Nosferatu is a signal:*** Scheel, D., Godfrey-Smith, P., and Lawrence, M. "Signal Use by Octopuses in Agonistic Interactions." *Current Biology* 26 (2016): 377–82.

240 ***this low posture just described:*** This work is being prepared for a forthcoming paper. Chancellor et al. In preparation. "*Octopus tetricus* Uses Stereotypical Displays in Interactions Affecting Access to Habitat Patches and Dens."

240 *a female gathered up a bolus of silt and shells, and directed it through the water at a male:* For a fuller account, see Godfrey Smith, P., Scheel, D., Chancellor, S., Linquist, S., Lawrence, M. 2022. "In the Line of Fire: Debris Throwing by Wild Octopuses." PLoS ONE 17(11): e0276482. https://doi.org/10.1371/journal.pone.0276482. For a news report, see Nature News 09 November 2022 https://www.nature.com/articles/d41586-022-03592-w.

242 *reach out with one arm to swat away a fish:* Sampaio, E., Costa Seco, M., Rosa, R., and Gingins, S. "Octopuses Punch Fishes during Collaborative Interspecific Hunting Events." *Ecology* 102, no. 3 (2020): e03266.

245 *harvest a thousand pounds of octopus in a morning, at one location:* Newman, M. A. "'Marijean' Octopus Expedition." *Vancouver Public Aquarium Newsletter* VII (1963): 1–8 (Vancouver Public Aquarium: Vancouver).

246 *this novel species:* For more about this species, see Hollenbeck, N., and Scheel, D. "Body Patterns of the Frilled Giant Pacific Octopus, a New Species of Octopus from Prince William Sound, AK." *American Malacological Bulletin* 35, no. 2 (2017): 134–44.

Hollenbeck, N., Scheel, D., Fowler, M., Sage, G. K., Toussaint, R. K., and Talbot, C. "Use of Swabs for Sampling Epithelial Cells for Molecular Genetics Analyses in *Enteroctopus.*" *American Malacological Bulletin* 35, no. 2 (2017): 145–57.

246 *more often found on silty slopes:* Benthic octopus species inhabiting primarily silty substrates often possess a longitudinal raised frill, flap, or ridge around the equator (lateral and posterior midline) of the mantle. In the frilled giant Pacific octopus, this appears almost as a continuous line of papillae that have merged (or nearly so). A few of these octopus species are adept at burrowing themselves in the silt. As yet, no one knows whether this Alaskan frilled giant Pacific octopus buries itself.

## ACKNOWLEDGMENTS

250 *I hope to honor the storytelling tradition:* This understanding of respectful use of Indigenous stories was informed by my conversations particularly with Apela Colorado, in asking and receiving permission from Simeon Kvashnikoff Jr. to relate portions of his story, and from the analyses of story work in Chapter 2 of:

Drabek, Alisha Susana. "Liitukut Sugpiat'stun (We Are Learning How to Be Real People): Exploring Kodiak Alutiiq Literature through Core Values" (PhD diss. University of Alaska Fairbanks, 2012).

# INDEX